Hope for a Heated Planet

Library of Congress Cataloging-in-Publication Data

Musil, Robert K., 1943–
Hope for a heated planet : how Americans are fighting global warming and building a better future / Robert K. Musil.
p. cm.
Includes bibliographical references and index.
ISBN 978-0-8135-4411-3 (hardcover : alk. paper)
1. Global warming. 2. Climatic changes—United States. 3. Global warming—Health aspects. 4. Carbon dioxide—Physiological effect. 5. Green movement. I. Title.
QC981.8.G56M876 2009
363.738′74—dc22

2008007762

A British Cataloging-in-Publication record for this book is available from the British Library.

Visit our Web site: http://rutgerspress.rutgers.edu

Text design by Karolina Harris

♲ This book is printed on recycled paper.

Manufactured in the United States of America

To Margaret Kirkland Musil (1915–2007)

Who taught me faith, hope, love

Contents

Preface

I am finishing this book in the final days of the George W. Bush administration. For those of us who love the environment and long for peace, it has been a dark time. Through it, I have often thought of the words of the poet Theodore Roethke that I first heard from scholar and activist Robert Jay Lifton: "In a dark time, the eye begins to see." The first light of dawn is now visible as the Democratic candidate, Senator Barack Obama, pushed by a growing grassroots movement, embraces action on climate change. To a lesser degree, though far more than President Bush, so does the Republican contender, Senator John McCain. That was far from the case when I began this project at the height of the president's wartime popularity.

This has turned out to be a book as much about hope and democracy as it is about global warming. Its central theme is that you and I can change history. What we believe, what actions we take, actually matter. It is an idea central to democracy. And it should give us hope. I disagree, strongly, with those who believe the American public has turned into a hopeless gaggle of consumers and couch potatoes who are content to let others rule their lives—or destroy the planet.

At the height of President Bush's popularity and influence it may have appeared that way. But national security and environmental degradation (especially global climate change) are complex, difficult, and abstract subjects. It has taken some time for us Americans to grasp the gravity of our situation, from melting ice caps to Iraq. This is especially true when our media mostly cover the White House and the Pentagon—regardless of the occupants—and report each utterance as gospel. Meanwhile, most of us are busy with jobs, families, and problems near home that we can actually see and do something about.

The result has been that global warming—caused by the vast outpourings of carbon dioxide (CO_2) and other pollutants from our cars, buildings, and factories—has increased and is picking up speed. But at the same time, so has a growing and revived environmental movement. It is joined now with new allies from the religious community, business, labor, medical and public health professionals, educators, and more. This new climate movement has deep roots in the big environmental groups, too often ignored or derided, that have been working to warn us and prevent global climate change since the elder George Bush's administration in the 1980s.

Their work is now bearing fruit. The public is becoming aroused and engaged. And, as a result, we will have a new, much more climate-friendly president and Congress in 2009. This book tells that story and also explains the basics of climate change and its effects on human health and well-being—not just on polar bears and penguins. But *Hope for a Heated Planet* is finally about solutions to our dilemma. I've tried to give you the best steps you can take, both personal and political, to make a difference and to get involved.

Like most authors, I like to imagine, of course, that our new president will take to heart every word I have poured out here. But even more important is that *you* do. My mother, Margaret Kirkland Musil, died after ninety-one wonderful years as I was writing. She taught me to love life, to love nature, to learn, to have faith, and to *act* on my beliefs. My first grandchild, Catherine Kirkland Unruh, was born shortly after. She will need the same lessons. So will all our children and grandchildren.

Global climate change, we now know, can be prevented by building a vibrant, healthy economy that does away with the belching furnaces, smokestacks, and combustion engines from the outmoded technologies of the nineteenth and twentieth centuries. But in a democracy, that will depend not on our new president, or the one after that. It is up to us. And the signs now are that many, many citizens and their organizations in this great nation are indeed aroused. But to prevail, we will need even more. That is really why I have written. I want you and your family and friends to join with me and millions of other Americans in making history. Nobody else can, or should, do it for you.

Family and friends have helped me throughout a lifetime of learning and activism, not only as I have been writing this book. They are too numerous to mention. Many are friends and colleagues from Physicians for Social Responsibility, from the Green Group of environmental organizations who brought us the Kyoto Protocol, and from other organizations I have happily collaborated with. Some are from American University, especially my graduate students in global climate change, while still others are scattered through churches, universities, medical centers, and communities across America where I have traveled, spoken, and organized.

But you would not be reading *Hope for a Heated Planet* if I had not been approached and encouraged to write by Audra Wolfe, then acquisitions editor for Rutgers University Press. My current editor, Doreen Valentine, a neuroscientist, kept out the worst purple passages and pushed me, properly, to rely on a bit more logic and a little less lyricism. We are all better off. My daughter Emily Kirkland Musil, now teaching at Trinity College in Connecticut, was finishing her dissertation for a PhD in history as I wrote. We shared similar agonies and ecstasies and were our own writers' group. My daughter Rebecca M. Unruh, a Washington lawyer, not only provided the inspiration of a granddaughter; she was supportive throughout, asked gently probing questions, and did not charge me billable hours. My sister, Marjorie Anne Musil, suffered a serious heart attack, survived to cheer me on, and has reminded me of the preciousness of life and the realities of health statistics that I try to make tangible in these pages. My wife, Caryn McTighe Musil, a scholar, leader, expert in higher education, and activist throughout her life has shared everything with me—from graduate school to grandchildren. Her determination to create a better world, and to have fun with yours truly while doing it, has been an inspiration since the days of miniskirts and mass mobilizations.

There you have it. This book will not solve global warming. You and I will have to do that. I believe we can and will. That is the hope I cling to. I've tried to explain everything I think you need to know clearly in one place. There are organizations, resources, and references throughout to take you even further on this journey once you start. But as a hero of my youth, John F. Kennedy, said: "A journey of a thousand miles begins with but a single step. Let us begin."

Robert Kirkland Musil
Bethesda, Maryland

Hope for a Heated Planet

Introduction

This book is an act of faith. It symbolizes hope. Holding it balanced on your knee in a favorite armchair or sneaking just a few more pages in bad light as your eyelids droop before surrendering to the propped pillows beneath your head, you and I are betting that there is a future, that human civilization will continue. We believe that ideas and people matter, that, in the words of Martin Luther King Jr., "the moral arc of the universe bends long, but it bends toward justice."

I agree with Thomas Jefferson, whose remark to John Adams, "I cannot live without books!" is stitched into a small, navy blue souvenir pillow on my mom's couch in the home where I grew up. I believe that books can change people's lives, as they have mine. They are, after all, one person talking quietly and intimately to another. And, of course, to thousands of others at a time, as if we were on a couch in the very same room. Radio, television, documentary films, and more modern media (iPods, Web sites, blogs, flash videos) may reach more people in attempts to touch them with words, pictures, symbols—the stuff of human culture. But only books are designed to be held close, to last, and to pull together in one place an extended, complex argument or idea. A book is to be savored, mulled over, reread, underlined, dog-eared and, ideally, to enter deeply into the heart, mind, and soul of the reader.

That is my hope for *Hope for a Heated Planet.* It is indeed an act of faith that you may come to share my belief, hard-earned over a lifetime, that together we can still save our planet, that we can help save our nation, our community, our families and friends from a dangerous, even deadly future. This book is after all (as you knew when you bought it) about global climate change. There are a number of very good books, articles, and films that already describe in detail how, if unchecked or unchanged, our collective appetite for fossil fuels—the oil, gas, and coal that power modern economies—will bring on "crisis" and "catastrophe." When the world's climate scientists put out their latest assessment in 2007, six years after their previous authoritative and fairly glum report near the beginning of President Bush's first term, the media mavens generally summed it up as "grim." And, of course, some scientists and environmentalists quickly quibbled that it was not grim enough. My gloomier colleagues may indeed be right about the fate of the earth. And they are surely right, in the tradition of the prophets, to try fervently to warn us what will happen if we play God with creation.

But I want to offer you hope. Not false or cheap hope, but realistic, hard-boiled hope as tangible as is the book you hold. To believe, as I do, that you and I can make a difference in the world, can create political change, can change the course of history, we will have to look squarely at the difficulties we confront with global climate change. Then before you despair, I want to share a number of things that should give you hope. And I want you to understand what my values are and why I write. Then you can judge for yourself, not only what facts and experience I share with you, but the messenger and his motivations as well.

If we citizens are to be more than mere spectators to the slow destruction of the planet, if we are to do more than assume that we are already cooked, or that some presidential, scientific, or technological savior will rescue us, we must understand for ourselves the science and the realities of climate change. If the problem is CO_2 emissions, we'll need to know how they work, where they come from, what parts of our daily and economic lives we need to change. And how can a bunch of scientists predict the future for all of us, nearly a hundred years from now? We need to know where our energy comes from, what the alternatives are, and how to change the policies that drive the decisions beneath all those investments in solar or wind or coal or nuclear power. But most important, I think, is that we need to believe that we are not alone in our concerns, that somebody is working to solve this problem, and that you and I can join in effectively.

To do this, we need to see global climate change as not just another issue. Disrupting the earth's sustaining climate is an immediate and central concern to our children and our communities, and to the creation that both the secular and the spiritual among us care about so deeply. As Al Gore said in a telling line from *An Inconvenient Truth:* "too many people move from denial to despair without stopping in between to figure out what to do." More and more people are stopping to look afresh at global climate change, far more than the small, remaining coterie of doubters and skeptics and those who profit from business as usual. I want to join with you and find that place between denial and despair where serious solutions and honest hope begin.

For me, it is having spent a lifetime in political movements for peace, the environment, and social justice that has given me hope. I do not despair over the common sense and decency of Americans. Social change comes slowly, sometimes maddeningly so. But however long it took, American citizens finally ended the war in Vietnam, the draft, segregation, the worst of the nuclear arms race, and the choking pollution that clouded our nation in the 1970s. To oppose the Iraq War and want to end it is now commonplace, even patriotic. I wish this had been so in 2003, when the war was launched, or again in 2004 when its initiator, George W. Bush, was reelected. But, ultimately, Americans have understood the harsh realities and the folly of preemptive war and an unwelcome occupation.

A similar phenomenon is happening, and has been for some time, with global climate change. Once the province of a hardy band of scientists, environmentalists, and a handful of perceptive reporters and politicians, global warming is now the stuff of news programs and of general conversation. How we got to this point is part of the story of this book. How we can keep going, who we can work with, and what choices you and I can make to turn climate change into a compelling personal and political issue is much of the rest. I have tried to keep things as simple and straightforward as possible while remaining faithful to the science and to the rapidly changing state of expert knowledge about global warming. That's because I believe deeply that in a democracy, decisions of life and death, whether for war, or for saving the environment and human life, should not be left to experts, to media pundits, or even to our elected representatives. You and I must decide. We must act. If we remain passive observers of great events as they play out on CNN or in the *New York Times,* we will have lost this nation's greatest legacy. Too often I am asked by friends or acquaintances: "What do you think will happen now in Congress?" or "Who do you think will win the election in November?" or "How bad is global climate change going to get?"

The answer, I say, is up to you and what you do about it. If you don't like the headlines, make your own. With climate change, I know it isn't easy. There really is a lot to know, to learn, and to judge. I have been at it for years. And I have written this book, in part, to learn more for myself and to share with you what I have found out. I have faith that armed with information from *Hope for a Heated Planet* and other sources, you will talk with friends, family, and neighbors. Then you will want to take action in your home, your place of worship, your community, and in the most open democratic political process on earth. You can help prevent global warming and the immense harm it has already started to do to the earth and our fellow humans alike.

But you must be saying that, like most authors, I am making some pretty big assumptions underlying these rather bold assertions. You deserve to know what they are. And in turn, because I hope to convince you and motivate you to act, I have found that most people respond best to honesty and openness. Throughout this book, I presume, there will be many times that you disagree with me, or doubt what I have to say, or suspect I have some hidden agenda. I cannot prevent that. But I do want you to know why I care so deeply about global climate change and why I believe something must and can be done about it.

Credo

I believe that the American republic and its institutions remain democratic and resilient enough to reflect the will of the people—however belatedly and imperfectly. I also believe that Americans and their organizations have the ability to create social movements that finally push and prod politicians to change policy even in the face of powerful and wealthy special interests or ingrained prejudices and misperceptions. I have seen the effectiveness of what the Quakers call "speaking truth to power." I have seen American leaders grow and change. Whatever the motivation, each human has the capacity to change and begin anew. It is an idea deeply embedded in both religious faith and in the democratic ideal. In the words of the old, beloved hymn "Amazing Grace," written by Captain John Newton, a British slave trader who renounced his traffic in human beings: "I once was blind, but now I see."

I also share a belief that we humans have a purpose and, yes, a moral obligation on this earth to protect and care for the goodness of all creation. In the biblical story of the Flood, both humans and animals are saved by two things. First, Noah (and Mrs. Noah!) show real ingenuity, pragmatism, and hope as they build an ark, shepherd all manner of strange creatures into it, and set sail for the future in the face of what surely looks like the end of the world. Second, there is God's promise, symbolized by a descending dove and a rainbow, that the world—Noah and all his descendants and the onboard menagerie and their offspring—will not be destroyed.

The actions taken by Noah and his wife represent a story of human hope that speaks across the centuries to progressive Christians like me, to Jews, to fundamentalist and evangelical Christians, to people of many faiths, and to purely secular and scientific souls as well. The key ingredient for each of us is that for the story to have a happy ending, for humanity and creation to go on, for God's promise to be kept, we humans, from Noah forward, have had to have faith and we have had to act. The fate of the earth and of all creatures both great and small (now known by ungainly terms like *endangered species* and *biodiversity)* depends on what we *humans* do. The Book of Deuteronomy sums it up pretty well. I have heard these words quoted by rabbis during the nuclear freeze campaign, by ministers in time of war, by physicians at conferences, and by an activist colleague and friend in the midst of tense climate change discussions in the White House and in Kyoto. "I put before you life and death. Choose life, that you and your children may live."

My own faith that we can act, change, and prevent the end of human existence was forged at the height of the Cold War when nuclear annihilation was a tangible and terrible threat. Norman Cousins, the editor of the influential *Saturday Review,* had devoted himself to nuclear disarmament from the moment the atomic bomb destroyed Hiroshima; he was moved to pen the editorial, "Modern Man Is Obsolete." Cousins went on to found SANE, the largest disarmament organization of the time, and developed relationships with both activists and statesmen around the world. In the spring of 1963, he visited with Soviet Premier Nikita Khrushchev, spending the day at his summer house or *dacha* on the Black Sea. Convinced that Khrushchev was open to a nuclear disarmament gesture from the United States, Cousins, upon his return, met privately with President John F. Kennedy. He urged him to halt nuclear testing in the atmosphere as the gesture to which he believed the Soviet leader would respond. It would take a leap of faith on Kennedy's part, however, because he felt that Khrushchev had previously humiliated, misjudged, and deceived him. And during the Cuban missile crisis, only months before, the two men had come perilously close to full-scale nuclear war. Indeed, on national television, at the start of the crisis, in announcing a naval blockade of Cuba, Kennedy had said chillingly: "We will not needlessly seek a thermonuclear war. The fruits of victory would be but ashes in our mouths. But neither will we shrink from it at any time it must be faced." Later, recognizing that he had come within a hair's breadth of destroying civilization, Kennedy, the consummate Cold Warrior, changed course. In his finest speech, the June 10, 1963, commencement address at American University, only a short distance from where I now teach about nuclear weapons and global climate change, the young president called an immediate halt to nuclear weapons testing in the atmosphere. He then went further, calling for an end to U.S.-Soviet confrontation and to the Cold War itself. Kennedy's rationale—combining Noah's ancient faith with modern understandings of the human and health consequences of nuclear radiation—was put with simple eloquence and in carefully constructed cadences. "We all love our children. We all breathe the same air. We are all mortal."

Within fifty-five days, the United States and the Soviet Union had signed the Limited Nuclear Test Ban Treaty. It banned all nuclear tests in the atmosphere—a practice that had spurred the nuclear arms race and that ultimately led to excess cancers and deaths around the world numbering in the millions. Kennedy (and his counsel and speechwriter Ted Sorensen) get the credit for this historic breakthrough. But it was built on the urgings of a respected peace leader and on a worldwide, more than decade-long nuclear disarmament movement composed of millions of citizens. They formed organizations, lobbied, carried out protests, and were led by a motley group of scientists, physicians, religious leaders and pacifists, worried mothers, Native American tribes, and statesmen from the nonaligned nations.[1]

A similarly hopeful turn of events occurred only two decades later. The nuclear arms race had not ended in 1963 with the end of testing. Both the United States and the Soviets had developed new and more dangerous nuclear weapons that could destroy not only civilization, but life on earth. A single missile carrying a MIRV (multiple independently targetable reentry vehicle), like the American MX or the Soviet SS-18, carried between ten and fourteen warheads, each with explosive power up to ten times that of the Hiroshima bomb. The net effect of a single missile strike on a metropolitan area would destroy up to 2,500 square miles and millions of people. A fleet of them would bring on "nuclear winter" in which the fallout, dust, debris, and toxins would block out the sun in the manner of multiple, huge volcanic eruptions from Krakatoa or the asteroid strike that ended the age of dinosaurs.[2]

When early in the administration of President Ronald Reagan, an avid hawk, anticommunist, and Cold Warrior, the United States announced that it was prepared to fight and win a nuclear war, the nuclear freeze movement, building on the remnants of the old disarmament and anti-testing campaign, was born. At its peak, the freeze won forty-four state and municipal elections and referendums calling for a halt to the nuclear arms race. The movement brought one million people out to New York City's Central Park for a disarmament march on April 12, 1982, and similar events occurred across Europe and other parts of the world. Both President Reagan and Soviet President Mikhail Gorbachev, we now know, were affected by this global shift in popular and political conditions.[3]

And so by October 1986, I witnessed Reagan and Gorbachev meet at a summit at Hofdie House in Reykjavik, Iceland, where the two men came tantalizingly close to agreeing to total nuclear disarmament. They ultimately did sign the INF, or Intermediate Nuclear Forces Treaty. It was the first nuclear treaty that literally eliminated and then destroyed an entire class of nuclear missiles.

The freeze movement, like the earlier disarmament movement from which it grew, had been initially ridiculed and denounced as a Soviet KGB plot by Reagan himself. The Reagan administration, before the freeze, had been building five new nuclear weapons per day, was heading toward 17,000 nuclear warheads, and announced that the United States was prepared to fight and win a nuclear war. Within ten years after Reykjavik, the entire nuclear weapons production complex was closed; strategic missiles were reduced; a Comprehensive Test Ban Treaty was signed; the Nuclear Non-Proliferation Treaty—through which more than 180 nations have renounced nuclear weapons—was extended permanently; and efforts to secure and get rid of nuclear weapons materials were under way. Indeed by 1996, before the decline of the Clinton administration and the advent of the George W. Bush administration, national governments, the United Nations, the World Court, and various religious and military leaders were all seriously discussing and proposing the complete *abolition* of nuclear weapons.[4]

It may not have been the proverbial activists' vision of world peace. But by the standards of human history, of American and world reform movements, and in changes in global attitudes—especially among the American public, policymakers, and the military—it was astonishing progress. Over my lifetime to go from the discovery of nuclear fission, to the destruction of Hiroshima and Nagasaki, to loving the Bomb during the days of *Dr. Strangelove,* to calls for abolition by the likes of Robert S. McNamara has been nothing short of a miracle. It has taken the threat of annihilation; sustained citizens' movements; the pronouncements of prophetic scientists, doctors, theologians, and nongovernmental leaders; and the near-conversion experiences of American presidents like John F. Kennedy and Ronald Reagan. But humanity has survived. Denunciations of nuclear weapons and their dangers are now commonplace.

So it is today with global climate change. The earth is literally facing the possibility of a flood of biblical proportions.[5] We humans must choose, act, build, and have hope and faith as never before. It is we who are risking God's creation. We alone can prevent that fate. And I believe we will.

A Public Health Approach

The structure of *Hope for a Heated Planet* roughly follows the outline of the graduate course on global warming that I teach at American University called "Global Climate Change: People, Power, and Politics." Given my background and experience in public health, communications, and public policy as head of Physicians for Social Responsibility, I approach the problem of global warming as a typical public health issue—like AIDS, tobacco, gun injuries, tuberculosis, automobile injuries and death—though an admittedly huge and complex one. The public health approach demands careful identification of problems and potential solutions using medical and scientific evidence, hard data if you will. It must be shown that the problem is real and is causing or may cause serious harm. Then you assess its roots in individual, social, cultural, and political factors. From this you can develop an action plan based on the realistic resources you can deploy. Next you have to change behaviors—individual, community, and public policy—to prevent and really fix things. You need to test the best ways to communicate with and persuade people to care enough, to change, and to take action. For a public health practitioner, it is never sufficient to merely study or research something that is harming large numbers of people. Preventing harm and taking responsible, effective action are what matter. These are, I hope, the sum, substance, and soul of this book.

In chapter 1, "Understanding Climate Change," I elaborate on what I mean by a public health approach to climate change and why I use it. Then I take you through the classic first step in solving a public health problem, looking at the scope, scientific basis, and danger from global climate change. Suffice to say that we are running out of time. If we do not level off the concentrations of CO_2 in the atmosphere at somewhere about 450 parts per million (ppm) by midcentury, thus keeping global temperature increases to about 2°F, we will ruin our civilization. The opening chapter offers a quick, and I hope readable, review of what the essential facts are, how we know what we do, and why the main questions about the reality and seriousness of global warming have been resolved. Given my approach, I also look more closely than do most global climate essays at the effects of climate change on human beings. Malaria, heat deaths, cardiovascular disease, asthma, and much more are all linked to the pollution that comes from burning oil, coal, and gas and are exacerbated by rising temperatures. I care deeply about glaciers, polar bears, and penguins. But human health is central to global warming. Unfortunately for attempts to bring about change in the United States, most of the devastating health effects related to climate change are happening in distant countries, often in the developing world. The suffering is very real, but media coverage is limited, and American voters usually put domestic and local concerns higher on their list of priorities. Thus, in chapter 2, "Home, Home on the Range," I look at global warming and its effects in the United States, especially in places that I know and love like my home in Maryland, in New York where I grew up, and in key states, like Texas, where I have visited often.

Chapter 3, "The Power of the Carbon Lobby," looks at the social and political causes of global warming, why the scientific debate has been so controversial, and why progress in tackling global warming has been so slow. In short, in public health we need to identify the forces we are up against so we can figure out how best to mount an effective response. Here you meet the oil, gas, coal, utility, and other interest groups that make up the carbon lobby. It has been fighting the environmental community for two decades to prevent any change.

Chapter 4, "Framing and Talking about Global Warming," looks at how best to understand, describe, and present our climate problem. In public health, how an issue is looked at, named, and what parts are focused on may be critical to success. Public health campaigns worked for years to get the public to understand and agree that smoking illnesses were more related to a powerful tobacco industry pushing an addictive substance than to the "weakness" of the individual smoker. The chapter addresses sharp critiques of the environmental movement by those like Michael Shellenberger and Ted Nordhaus in "The Death of Environmentalism." They have argued that our lack of progress results from the failure of the nation's biggest environmental organizations to mount bold appeals, effective campaigns, or to get beyond cautious Capitol Hill incrementalism. These critics believe that the very term "the environment" is a problem. Moreover, they say that big green organizations have failed to excite constituencies beyond hard-core environmentalists or to engage the values and passions of ordinary Americans.[6] I agree with some of this criticism. But I argue that the big green groups had already changed well before the latest blasts. They have been remarkably effective, especially given the many obstacles to progress. For years, the big greens have been reframing climate change as a human health issue, as a threat to national security, and as an opportunity to find energy solutions that can revive American communities and our economy. Here I also show that modern medicine and microbiology allow us to understand that the environment is not just a thing outside us. It is within us. Scientific studies show that even tiny amounts of pollutants travel the globe and end up inside our bodies and our cells. That's why I examine carefully the effectiveness of environmental health campaigns against mercury and air pollution that stalemated the most reactionary forces during the Bush years. These victories belie the charge that environmental groups have neglected grassroots efforts and bold calls to action.

In chapter 5, "Assessing the Big Greens," in order to set the record straight and to decide whether Washington green groups should be the core of a renewed climate change movement or be simply left behind, I offer my best judgment of the mainstream national environmental organizations in light of their most serious critics. In our public health assessment, we need to determine whether a social movement that can blunt the power of the carbon lobby and arouse the nation is possible. I give a short history of what big greens have actually done to make climate change an issue, especially in the Clinton years, when they pushed and prodded to get the Kyoto Protocol while holding off most of the reactionary attempts in the Congress to undo environmental gains. In the process, they sowed the seeds of a "new" climate change movement.

There is wide agreement even among traditional big green organizations that if we are to really stop global warming, the United States will need a stronger and wider environmental movement with more allies and new constituencies than we have seen so far. Chapter 6, "The New Climate Movement," looks at this emerging phenomenon and the efforts of big green groups and grassroots activists alike to stimulate it. Here I focus on the remarkable and growing shift to real concern about climate change among businesses, academics, youth, and many others. And I pay particular attention to the burgeoning faith-based climate movement. It represents one of the biggest gains for the climate change movement and tremendous hope for reaching heartland America.

With our climate change dilemma carefully documented, the social causes weighed and evaluated, and an assessment of how we might mobilize and communicate, I turn in chapters 7 and 8 ("Where Do Emissions and Energy Come From?" and "Energy Futures") to look at the possibilities of prevention—first by ridding the world of fossil fuels: oil, gas, and coal. (I also look closely at nuclear power, as some experts and even a few environmentalists have argued that we need it. I disagree.) Then I discuss hopeful trends in getting our energy from clean, renewable, and healthy sources like wind power and solar energy, new fuels and technologies for cars and other vehicles, efforts to contain sprawl, and the growing revolution in sustainable "green" design for buildings. I also highlight in-depth analyses that show, contrary to conventional wisdom, that we can get enough energy from renewables while cutting carbon emissions to avoid the worst effects from global warming.

The last two chapters are the final step in any public health campaign—action to prevent and eradicate a major health problem. They are also designed to help you get involved. I outline practical, personal steps you can take in your own home and community in chapter 9, "Creating Hope." And I include a brief guide to organizational and political engagement in the final chapter, which repeats the title of the whole book. It is a primer on policy changes you should demand from your representatives, especially of candidates for office, and from our new president.

There is a lot to consider in *Hope for a Heated Planet.* But there is no larger or more important issue before the American public or the world than climate change. I really do have faith that we will soon make rapid progress in battling global warming, that rapid change is possible in our public consciousness and in our public policy. But, whether or not that actually happens, Mr. and Mrs. Noah, is really up to us—to you and to me.

1

Understanding Climate Change: A Public Health Approach

Toward the end of his final term in late 2007, President George W. Bush called, belatedly, for an international meeting to discuss global climate change. Then in 2008, at the international negotiations in Bali, Indonesia, designed to set the parameters for a new climate treaty to update and replace the Kyoto Protocol, his administration, though not joining the treaty, no longer tried to block progress by other nations. These were late, futile gestures from a leader whose political stock had been plummeting for some time. Ignored for some seven years, global climate change had burst into view again between 2005 and 2007 for a number of reasons: the release of a major scientific report that the Arctic was melting;"[1] the unprecedented natural devastation on U.S. soil by Hurricane Katrina; two more years of record of heat in 2005 and 2006 followed by an ongoing severe drought in 2007;[2] the release in early 2007 of the Intergovernmental Panel on Climate Change (IPCC)'s Fourth Assessment of climate change;[3] and, of course, the sweep of the Republican Congress that had featured climate change deniers like Senator James Inhofe (R-OK). By 2008, global warming was mentioned, along with some sort of plan to deal with it, by almost all the leading presidential candidates.

But why this seemingly sudden shift in public attitudes? A single movie from Al Gore? Some convincing new piece of scientific evidence? A change of heart from the corporate leaders who profit from producing polluting fossil fuels? Some of these factors provide partial answers. Even more are involved in a complex story of social change. But I believe that global climate change and energy policy have surfaced as matters of serious public interest because of a dedicated, rapidly expanding global climate change movement. It was begun by traditional environmental organizations that brought us the Kyoto Protocol in 1997. They have over many years fought an often lonely, though effective, fight. But now there are new allies.

Climate change will surely sink again beneath the waves of public consciousness and media attention. The war in Iraq, which derailed the president's second term, remains a major concern for Americans. So does a deteriorating economy. But climate change is now with us. It will not go away. Each time it reappears, I am convinced that more Americans than ever must understand and deal with it. Such a task will require a more positive vision of the future from environmentalists and from our leaders. And surely an even broader climate change movement that addresses greater numbers of ordinary Americans will be essential.

As citizens, you and I need to know the facts and make our own decisions. But, most of all, we need to believe that something can be done. As we have seen with the war in Iraq, changing understanding, attitudes, behavior, and political preferences takes time. But it can be done. Throughout this book, drawing on my own experiences, I want to look at global climate change, as I have at war, as a typical, if huge, public health problem.

My approach reflects my background as an activist, teacher, and engaged journalist; my training in public health; and my years at Physicians for Social Responsibility (PSR). At PSR I learned that a combination of moral passion—the stuff of activism, preaching, and teaching—works best when yoked to the credibility of hard data. You need a hypothesis rooted in scientific analysis. There must be a demonstrable correlation between the problem described and the solutions offered. You also should rely on some ancient verities: an ounce of prevention is worth a pound of cure; a stitch in time saves nine; and, the admonition at the core of all medicine and healing, since Hippocrates: "First, do no harm."[4] Public health finds the causes and solutions to health problems that affect large populations. The cures of clinical medicine or the steps taken solely by individuals are not enough. Only prevention—getting to the root of the problem and bringing about significant changes in attitudes, behavior, social structures, and public policy—will ultimately work. The public health approach must make sure there is sufficient measurement, data, biological plausibility, and real or potential human harm to support the diagnosis of a major health concern. Then, once the causal factors are assessed and determined, an action plan is designed and carried out. With human life and human society at stake, simply understanding a crisis is insufficient. The ethical thing is to take action.[5]

My favorite story in the history of public health is the classic tale of

Dr. John Snow, a British physician in the Victorian era. Snow correctly explained and stopped a cholera outbreak in London when such horrifying epidemics were common in the West, as they are still in the developing world. In the time of Dickens and Darwin, doctors and scientists believed that cholera was related to "miasma"—dank vapors in the air. The offending bacteria, *Vibrio cholerae,* had not yet been identified. But Snow correctly believed that the disease was carried by polluted water, not air. He methodically enumerated and mapped the cases of cholera on London street grids. The correlations showed that the outbreaks could be traced to water from a single source. It was the public water pump on Broad Street. Snow convinced the city powers to shut it down. Shortly thereafter, the cholera outbreak ended completely. The combination of good science and data, medical and biological plausibility, tracing causes, and taking action remains the heart of public health. It is why PSR honors its outstanding activists with a Broad Street Pump Award.[6] If the American public, you and I, are to take effective action based on our deepest values about humanity and the planet, we had better first look carefully at the science—the causes and effects of climate disruption. Then we can develop an appropriate plan.

The State of the Science

The debate over the science of global climate change is largely over. Clear and obvious warnings about the dangers of rising CO_2 levels and temperature were heard as early as the 1960s. By the time of the Carter administration—after a National Academy of Sciences report and a presidential panel validated global warming projections—the American government was inclined to take action. But between the growing unpopularity of Carter and the concerted forces set to oppose any action, there was little realistic hope that his administration could create much change in energy or climate policy. As in our day, the immediate national security "crisis" of American hostages in Iran then determined the 1980 election framing, polls, and outcome.[7] We had tightened our oil belts a notch. But we then quickly returned to business as usual. This despite reports on the need to take action on climate and CO_2 emissions in the '80s and throughout the '90s. By then thousands of international experts had reached consensus that climate change was real.

But it was not until 2001, after Kyoto, that the Intergovernmental Panel on Climate Change (IPCC), which had been issuing in-depth reports and warnings, finally concluded that global warming is almost certainly caused by human activity.[8] Spencer Weart, in his account of how we came to know what we do, *The Discovery of Global Warming,* describes how for decades climatology and other sciences lurched forward by fits and starts, following false leads, rejecting theories that did not pan out, and improving computer modeling and the collection of data. Each criticism was examined, each contrarian idea put to the test in the lab and in the field. The conclusion? The search is over, climate change is here.[9] Then, throughout the Bush years, more scientific evidence emerged. It was finally assembled again in 2007 as *Climate Change 2007: The Physical Science Basis, Summary for Policy Makers,* one of many different sections of the Fourth Assessment Report of the IPCC. The news was released in Paris on February 2, 2007. The predictions in the report confirmed once more that climate change was certainly caused by our use of fossil fuels. The IPCC also made clear for the first time that even worse climate scenarios or "surprises" were possible. These gloomy forecasts are based on nineteen separate computer models, about twice as many as in the past. They draw on recent peer-reviewed studies that show that temperatures are the hottest in thousands of years.[10]

If the science of global warming has been so convincing, why has the public been seemingly so slow to understand it? Part of the difficulty for the ordinary person has been making sense of isolated, brief news reports. They are what scientists call "anecdotal evidence." In the world of the newsroom, even dramatic events are reported as single, unconnected catastrophes. Or there is not sufficient evidence to assert *without doubt* that a single incident is related to or caused by climate change. The result? No big picture; only isolated, passing concern. Consider the string of fairly startling events related to global climate change just since the 1990s.

In his 1997 book on climate change, *The Heat Is On,* journalist Ross Gelbspan begins with the stunning, unforgettable collapse of part of the Larsen B ice shelf in Antarctica. A huge mass of ice the size of Rhode Island breaks off and plunges into the Weddell Sea.[11] By 2003, another huge portion of Arctic ice, long thought to be less vulnerable to warming, breaks off roaring into the sea. Gelbspan faithfully records this episode and the three huge Antarctic ice disintegrations since 1995 in his second climate book, *Boiling Point.*[12] Other reports, news items, and authors tell us that millions have died in record floods in China, India, and Bangladesh; that massive destruction has occurred in Central America as a result of Hurricane Mitch in Honduras; that cholera outbreaks in South America are related to warm water and increased El Niño effects. Record ice storms hit New England, and heat waves and fires grip the American West, while more than fifty thousand people die in Europe in 2003 as a result of record heat.[13] The stories pick up pace as climate effects become more obvious. We read of the extinction of frogs in the rain forest,[14] polar bears drowning as they hunt for seals amidst thinning ice and open water in the Arctic,[15] and, finally, the most devastating hurricane in American history, Katrina, roars through and destroys much of a Gulf Coast area the size of England and inundates the city of New Orleans. Nearly 2,000 Americans are killed; a million people are forced from their homes, with 375,000 still refugees at the end of the summer of 2006. All this scary stuff is indeed related to climate change.[16] But it is changing patterns over long periods of time that must be charted and seen. The dots of individual disasters must be connected.

Stripped to its essentials, the science of global climate change can be learned by nonspecialists. But climate science, like other science, is often intertwined with politics. Nonscientific values shape what questions are asked, funded, and accepted. In the United States, measuring changes in the atmosphere, influencing weather patterns, and government investment in meteorology were fueled by the Cold War. We sought to monitor Soviet A-bomb tests, to seed the clouds to impede enemy troop movements during warfare, and to better guide war planes and missiles. U.S. government support for science grew rapidly, as did the opportunity to get grants for other interesting work.[17] By 1957, Roger Revelle (later Al Gore's teacher at Harvard) was able to get funding to measure CO_2 concentrations in the atmosphere. He hired Charles David Keeling, a young scientist and engineer, to do it. Keeling began measuring CO_2 on Mauna Loa in Hawaii and at a remote spot in Antarctica. After initial contamination of samples, serious results began in 1958. They were published, with graphs, in 1960.[18] The now fairly well-known science graph, the Keeling curve was born. It shows jagged lines of annual CO_2 concentrations, as actually measured, moving steadily upward. Now through analyzing bubbles trapped in ice, scientists have been able to accurately estimate temperatures correlated to CO_2 concentrations as far back as 650,000 years. What the air samples at Mauna Loa and from ancient bubbles trapped in Antarctic ice show, when temperature lines are then superimposed, is pretty simple. There is a clear correlation between increasing amounts of CO_2 in the atmosphere and rising average temperatures on the earth. Scientists then project the graph lines into the future. As memorably captured in one of the best scenes in *An Inconvenient Truth,* the two lines, CO_2 concentration and temperature rise, both begin to rise steeply. Finally, they take off almost vertically, causing Al Gore to have to mount a cherry picker to point at them rather comically as he is hoisted up far above his audience. The result? Early theories and the first measurements of rising global temperatures have all now been confirmed by a half century of meticulous data collected high atop a volcano in Hawaii, from ice cores drilled two miles below the earth's surface, and from other measurements.[19]

The physical explanation for the warming is fairly simple. The temperature of Earth depends on energy from the sun. Some solar energy is reflected back to space. The rest is blocked and trapped by the atmosphere. Its thickness and composition depends on Earth's size and, hence, its gravitational pull. Our sun, the star that provides our solar system and Earth its heat and light, gives off radiation of varying wavelengths as hydrogen gas is fused by immense gravitational pressure. If we were closer to the sun or had no helpful atmosphere, we would be broiling like Mercury. No life could be supported. If we were further away, like Neptune, things would be frigid and lifeless. Our atmosphere on Earth is composed of 78 percent nitrogen, 21 percent oxygen, and 1 percent other gases, most of which are "greenhouse" gases, meaning that they trap heat. The main ones are CO_2, whose concentration has been between .03 and .04 percent and is now rising; and water vapor, which varies between 0 and 2 percent. In his *Global Warming: A Very Short Introduction,* Mark Maslin reports that without the greenhouse effect of these two gases, Earth's average temperature would be 4°F (−20°C) instead of our balmy average in the high 50s.[20]

Simple so far? Then why has there been such controversy over this basic point about the correlation of CO_2 and temperature? Climate skeptics had pointed loudly to what appeared to be discrepancies between actual recorded satellite measurements of temperature in the upper atmosphere versus temperatures measured on the earth's surface. The satellite temperatures were a bit cooler than those of the surface. But climate models (computer simulations and projections) and atmospheric science say it should have been the other way around. As it turns out, scientists have to adjust findings from satellites based on the exact time they cross over parts of the earth. Satellites show subtle shifts in their orbits caused by the earth's "wobble." Our not-perfectly-round planet spins a bit off-center. When adjustments and new calculations were made, the measurements, the science, and the models were all in alignment. Atmospheric temperatures at the surface and in the upper atmosphere agree. They show warming. Even the original skeptics from the University of Alabama were convinced.[21]

As if to wipe away any lingering doubts about the correlation between temperature and rising CO_2 levels, my colleague, Dan Lashof, a senior scientist at the Natural Resources Defense Council (NRDC), points out that two independent research teams of scientists at the Scripps Institute for Oceanography and at NASA's Goddard Institute for Space Studies reached similar conclusions. Our vast oceans are now heating up to a depth of more than 1,000 feet. No ambiguities in measurements here. As Jim Hansen, Director of the NASA Goddard Institute and a lead author put it: "This energy imbalance is the "smoking gun we have been looking for."[22] With the release of the IPCC Fourth Assessment in 2007, Andrew Weaver, a coauthor of the study, put it even more bluntly: "This isn't a smoking gun; climate change is a battalion of intergalactic smoking missiles."[23]

There is no serious disagreement today that the earth's surface, atmospheric, and ocean temperatures are rising in lock-step with the growing emissions of CO_2 that have occurred since the Industrial Revolution, especially in the late twentieth century.[24] So we have global warming. How does that affect global climate and the familiar patterns across the globe that humans have gotten used to? Two of the main factors in changing our climate are changes in the hydrologic cycle, or how and where water and rain go and the proportion of ice, ocean, and land mass that exist on Earth. Let's look at water.

As the earth warms, more of the water on earth—in the form of ponds, lakes, streams, rivers, and oceans—evaporates. It is held in the atmosphere as water vapor. We get moister, more humid air. Two problems here. The first is that water vapor is itself a greenhouse gas, little discussed as yet. It traps heat and creates a feedback loop. More water evaporates, more moisture traps more heat, and so on. Next, the water vapor in the air and clouds falls as rain, snow, sleet, hail. And it does, and will continue to do so, in increasingly uneven ways. Much of the United States is moister and warmer than in past years with both more heat waves and more rain. But in some parts, there is very heavy rain and flooding; in others, drought and parched earth. In sum, the patterns of rainfall and weather events across the planet are becoming disrupted, variable, and more extreme. Africa, Australia, and the American West undergo drought, while other areas suffer heavy storms and rain.[25]

But perhaps the most important, dramatic, and now visible shift in climate is occurring with the ice on planet Earth. There is a large and growing literature on ice and climate. I have been drawn to it because I despise hot and humid weather. I have long sought out mountains, glaciers, and cool climates of all kinds. One of my fondest childhood memories is when I first got to the Rocky Mountains on a family vacation to Colorado from Long Island and was able to throw snowballs in July 1952. July snow in the mountains was so much fun, so weird, and memorable that I did the same thing with my wife and two daughters twenty years later on the Olympic Peninsula in Washington State. I recall being extremely happy that summer when I sighted my first glacier in Glacier National Park.

Since then, glaciers have been retreating ever more rapidly in Glacier National Park. It glistened with more than 150 glaciers when naturalists hiked it in the nineteenth century. Now only thirty-five remain. Scientists predict that they will all be gone in about twenty years.[26] Similar glacial retreat is occurring around the world. The uncovering of the frozen remains of a five thousand-year-old Bronze Age man in the melting Ötztal glacier in the Tyrolean Alps only confirmed my fears. Ötzie, as he was dubbed, was found near the border of Italy and Austria not too far from where my paternal grandfather came from.[27] All this rapidly dwindling ice unfortunately makes for stunning and shocking side-by-side photos when old postcards are set next to contemporary views from Austria, Switzerland, the Rockies, South America, and other formerly icy spots.[28]

But the real danger comes from news that the ice covering Antarctica and Greenland is now melting at faster rates than previously thought. If unchecked, these changes could lead to truly catastrophic rises in sea level. Until recently, mainstream scientists assumed that the Greenland ice covering was not likely to melt away. Now a number of reports indicate that Antarctica and Greenland are melting at accelerating rates. The problem has been summed up by NASA's Jim Hansen, who explains that "Ice sheets waxed and waned as the Earth cooled and warmed over the past 500,000 years. During the coldest ice ages, the Earth's average temperature was about ten degrees Fahrenheit colder than today. So much water was locked in the largest ice sheet, more than a mile thick and covering most of Canada and northern parts of the United States, that the sea level was 400 feet lower than today." During warmer interglacial periods (about 2°F warmer than now, Hansen tells us) "sea level was as much as sixteen feet higher." If present trends, or business as usual, continue, temperatures will rise another 5°F in this century. Sea level could rise *80* feet or higher—a level unseen in 3 million years. That would create a deluge that would erase most East Coast cities—Boston, New York, Philadelphia, Washington, and Miami. It would swallow whole much of the state of Florida. Some 50 million people in the United States live below that 80-foot sea level. Meanwhile, China would have 250 million displaced persons; Bangladesh, 120 million (almost its entire population); and India another 150 million.[29]

Such a flood, of biblical proportions, is now a credible scientific possibility. The alternative, as Hansen and other scientists point out, is to make changes in our energy and economic patterns now. We need to keep concentrations of CO_2 below about 450 ppm and limit future temperature rise to under 2°F. If we do that, we would then likely see ocean rises of about 2–3 feet in this century. But there would be time to adapt and work on further prevention.

But as our scientific understanding of global warming advances, even the time frame in which climate shifts might occur has been called into question. Until fairly recently, scientists had been convinced that major changes in global climate took eons to develop. The contemporary scientific consensus had been that even though humanity is altering the climate, real transformations would likely take centuries. But more recent research has shown otherwise. Huge alterations in climate have occurred in the past with startling suddenness, sometimes measured in mere decades. Although the exact mechanisms are debated, we now know that the onset of climate periods such as the Younger Dryas came swiftly. What are the implications for the future of climate and our complex global civilization that depends on a fairly steady climate? We may be poking at a climate hornet's nest with a very sharp stick. The Pentagon set off a brief media frenzy in 2004 by authorizing a study of the implications for U.S. national security if the thermohaline conveyor—a huge, circulating oceanic current that warms northern Europe—were to suddenly shut down. The report speculates on some worst-case scenarios if continued Arctic melting of fresh water altered the salinity and density of the ocean near Greenland. That would divert the huge flow of warm waters via the Gulf Stream and other oceanic "conveyor belts" that flow toward Europe. As a new Ice Age engulfed NATO nations and disrupted agriculture and normal shipping lanes, the result would be chaos and fighting over oil supplies, arable land, and other human resources. Such a scenario is considered extremely unlikely. But it and other rapid climate "surprises" are part of the grimmer realities announced in the 2007 Fourth Climate Assessment.[30]

More likely are rapid shifts in some regional climates. These are more sensitive to change than entire global systems. Such a sudden transformation into an arid climate brought on the end of one of the most flourishing ancient civilizations. John D. Cox, author of *Climate Crash,* recounts the early case of the Middle Eastern Akkadian Empire. It suffered a sudden, climate-related drought in 2200 BC that lasted for three hundred years. It destroyed a far-ranging realm ruled by Sargon the Akkadian. The rapidity of this civilization's demise was first discovered by archeologist Harvey Weiss in the 1980s as he excavated the Tell Leilan site in northern Mesopotamia (modern Iraq). Humans had been cultivating barley and wheat on the Habur Plains there since Neolithic times some eight thousand years ago. Further research corroborated the dates of both the onset of dry, windy conditions and the social chaos that followed, despite sophisticated efforts by Akkadians to adapt and build grain storage and water regulation technologies.[31] What scientists call climate "surprises" may be extreme worst case and highly unlikely scenarios. But they cannot and should not be ruled out.

Climate Science, Humanity, and Health

The science of climate change is clear, as are the major effects of global warming on the planet—including rising CO_2 levels and temperature, melting ice, rising seas, extreme weather events, shifting global patterns, and more. But it is still not immediately apparent how these phenomenon affect human beings—people like you and me. I studied climate science and its adverse health effects throughout the '90s and spoke about it to delegates and the press in Kyoto. But it was pretty abstract. The truly disastrous potential of tinkering with our earthly home struck me most forcefully when I attended the 2002 World Summit on Sustainable Development (WSSD) in Johannesburg, South Africa. Climate change was already starting to exacerbate conditions for people throughout southern Africa.

If Americans remember much at all about the 2002 World Summit, it may be that then Secretary of State Colin Powell, widely respected and perceived as a thoughtful, moderate brake on the Bush administration, was booed loud and long as he spoke to assembled international delegates and representatives of nongovernmental organizations (NGOs) from around the world.[32]

But what seemed to me the real news at that huge, sixty thousand–person summit were the pleas of various government ministers throughout southern Africa. Because Physicians for Social Responsibility is concerned with global climate change, children's environmental health, and issues of global conflict, I met and talked with ministers of health, environment, and foreign affairs. Their message was invariably the same. This conference and all the major powers are ignoring genocide in southern Africa. Some 15 million people are likely to die from combinations of drought, hunger, AIDS, malaria, childhood diarrheal illnesses, and increased conflict over scarce resources. Disrupted climate only makes things worse.

I had first seen children sick and dying of diarrhea in refugee camps in Central America twenty years before. I was sickened and shocked. Like a lot of Americans, I had thought of diarrhea as an embarrassing minor intestinal upset related to overeating or the flu. I did not realize that 1–2 million children under the age of five die each year from it.[33] Or that the waterborne parasites and bacteria that lead to it breed more easily as temperatures rise. And so the image of dying children stuck with me. So has the horrifyingly ironic sight of lush, green sugar plantations as I traveled in neighboring Swaziland before the 2002 Johannesburg summit. The fields were steadily irrigated in the midst of drought in order to produce sugar for foreign markets. Meanwhile Swazi women walked the dusty roads carrying large plastic jugs of water on their heads. Their children gathered sticks and sucked on the chopped remnants of cane that littered the roadsides.

Africa remains in crisis today. Climate change is already causing some of its most devastating effects on the people of the sub-Saharan tropics. But because of the nature of the debate over climate change, the children of South Africa or Swaziland are rarely mentioned as victims of our addiction to fossil fuels. But increases in temperature in sub-Saharan Africa from the IPCC mid-level projection of warming could increase global heat deaths sixfold, worsen droughts, and exacerbate problems with agriculture and hunger.[34]

Even the current wave of interest in climate change looks mainly at the increasingly obvious and sometimes spectacular geophysical effects of global warming. The most vivid images remain those of crashing and melting ice in the Arctic and Antarctic, the drying up of the Aral Sea, or side-by-side photos of places familiar from literature and film, like Mt. Kilimanjaro, which now shows more barren, dark rock on its peak than glistening snow and ice. The progression for climate chroniclers seems to be to calculate the changes in atmospheric and ocean temperatures, then explain or show the geophysical effects, then the impact on wildlife and ecosystems, and then human settlements. Only at the end of this logical chain do climate analysts assert fairly generally that these changes may also be bad for human beings and society. Because most people understand that humans have adapted to and lived in climates ranging from the arid, broiling outback of Australia to the snow and ice of the Arctic Circle, the conclusion too often may be: "well, we've adapted before and live in all sorts of climates, what's the big deal if the earth's climate does change?"

Humans have evolved over a very long period of time. But the organized societies that sustain modern humans have only been created as we and other species adapted to the relatively stable climate following the most recent Ice Age. A rise in global temperatures of just about 9°F permitted the establishment of agriculture and permanent settlements. Then with the advent of public health improvements and modern medicine in the late nineteenth and early twentieth centuries, especially in wealthy nations like the United States and Great Britain, life expectancy rose rapidly. Many infectious diseases were eradicated; the outlook for the future of human health looked quite promising. Yet human health and diseases have changed and evolved along with changes in the environment. Leading climate and health researcher Dr. Tony McMichael has reminded us that over the course of human evolution and the development of human society, we have been far more affected by climate, environmental, and social factors than by individual circumstances or medical advances. Indeed, just as human health was improving in the twentieth century, a series of developments linked to industrialization, urbanization, and ecological and climate change have begun to threaten human health and offer uncertain futures. Mechanization has led to inactivity and obesity in the developed world, while the spread of toxic chemicals and various industrial processes has harmed human health worldwide. With longer life and the stresses and exposures of modern living have come cancers, increased heart failure, and other chronic diseases. And as human activity disrupts climate and various ecosystems, human health will be affected in a number of ways.[35]

The clear links between changes in climate and the environment or ecosystems upon which human health and well-being depends are not yet commonly discussed or understood. But even when the adverse effects of climate change on human health around the globe are presented to Americans, the worst calamities seem borne by people in distant, developing nations. Their lives are often unfamiliar, little understood, and still rarely covered in the media. Sea-level rise, for example, is predicted under mid-level estimates by the IPCC, to be about 20 inches by the end of this century. It will cause substantial disruption of coastal regions and small island nations.[36] But far worse would be the effects, especially in the developing world, of the high-end IPCC estimate of 1 meter, or 39 inches. That would lead to some 50 million new refugees, the flooding of 5 million square kilometers, and the destruction of one-third of world croplands.[37]

The same is true for the spread of so-called vector-borne (spread by insects) diseases. According to the IPCC, there will likely be increases in malaria, dengue fever, cholera, schistosomiasis, and encephalitis. But estimating potential disease outbreaks remains tricky. It is subject to criticisms from climate skeptics and the built-in caution and conservatism of the scientific establishment. A careful look at a critical disease like malaria can offer a good idea of why warning about the profound effects of climate on human health is complicated and has sometimes failed to gain sufficient attention.

One of the leading experts on the correlations between malaria and climate change is Paul Epstein, associate director of the Harvard Medical School Center for Health and the Global Environment. A colleague and friend of mine, Epstein is a wiry, energetic, and deeply committed physician with a soft, sincere, and compelling voice. He is one of the relatively few doctors willing to write and speak not only to his peers, but in popular forums or films. His efforts to do so have drawn some criticism that exemplifies the difficulty of taking complicated medical and scientific subjects to nonscientific audiences. Because malaria is both preventable and treatable, its outbreak depends on a number of public health factors: farming methods and irrigation, mosquito control, surveillance of new cases, the susceptibility of victims, the treatment of those with the disease so it does not spread, and so on. The proper climate and temperatures for breeding and biting is necessary. But there is no *single* cause for a malaria outbreak. Nevertheless, what is certain is that mosquitoes only flourish within a warm, moist climate and that such zones are spreading. The most deadly form of malaria is the parasite *Plasmodium falciparum.* It is the primary killer in Africa. For the disease to occur and spread, one of several species of *Anopheles* mosquitoes, usually *Anopheles gambiae,* must bite an infected person, then bite someone without malaria, injecting the parasite into the bloodstream of the second human host. As temperatures warm, the *Plasmodium* parasite matures more quickly. At 68°F, it incubates within twenty-six days. At 77°F, it takes only thirteen days. Similarly, when it is warmer, mosquitoes feed and bite more frequently.[38] Because the body temperature of mosquitoes is directly affected by surrounding temperatures, if they stay below about 60°F, then malaria parasites cannot mature inside their hosts at all.[39]

In developed nations like the United States, which had malaria zones until the 1930s, our advanced public health infrastructure means that a few malaria cases are less likely to turn into an epidemic. But in Africa, and other underdeveloped areas in Asia and Latin America, all things being equal, the spread of warmer areas conducive to mosquitoes means the spread of malaria. So as attention slowly turned to questions of climate and human health after the 1995 IPCC report, Epstein and others noted that malaria had begun to appear at higher levels in the mountains of Kenya. With climate change, the cool temperatures that kill off mosquitoes or hinder their breeding began to disappear at higher and higher elevations. The conclusion drawn by Epstein and others was straightforward. Continued temperature increases would spread the areas where mosquitoes could flourish and malaria spread ("Malaria endemic zones").[40] Epstein then wrote an article in *Scientific American* and began to have many more media appearances.[41] Immediately, both climate skeptics and particularly cautious scientists began to poke at the Kenya case. They noted that additional factors were or might be at work. I filled in for Epstein on a cable TV panel at one point with fellow environmental scientists. But each participant was reluctant to state in a direct, understandable way the obvious fact. In Africa, where most malaria occurs, rising temperatures would ultimately mean more malaria.

Evidence of malaria's spread and its relation to climate change continues to accumulate. Malaria is indeed far more prevalent in higher elevations in Kenya and in the capital Nairobi itself.[42] A careful, controlled study of malaria in Debre Zeit in the highlands of Ethiopia tracked malaria and temperature changes from 1986 to 1993.[43] Controlling for other factors, the association between warming trends and malaria, in the words of Dr. Jonathan Patz, principal lead author of the Second and Third IPCC Assessment Report sections on human health, "could not be explained by drug resistance, population migration or level of vector-control efforts."[44]

Currently, some 40 percent of the world's population is at risk of getting malaria. Some 75 percent of cases occur in Africa. Of those with malaria, three-quarters are children. According to the World Health Organization (WHO) more than three thousand kids die from it every single day. All told, the WHO has estimated that 300–500 million cases of malaria occur every year. Estimates of the deaths caused by it range from 1 million to 2 million per year. By 2050, estimates are that 45–60 percent of the world's population could live in malaria-endemic zones, with new cases and deaths far outpacing current levels.[45] But as Epstein reports in his most recent comprehensive study of climate and health, "the true impact on mortality may be double these figures if the indirect effects of malaria are included. These include malaria-related anemia, hypoglycemia (low blood sugar), respiratory distress and low birth weight." What this adds up to is that one in nine of all years of life lost in sub-Saharan Africa is attributable to malaria.[46]

In addition to warming temperatures, the spread or outbreaks of malaria can also be related to flooding, heavy rains, or drought. All these are associated with growing climate disruption. With flooding, far greater areas of pooled water for mosquito breeding appear. With droughts, human populations often flee to areas with more water and—malaria. In 1998, after the 6 feet of rain dumped onto Honduras during Hurricane Mitch caused flooding, malaria and other diseases flourished. Mostly unnoticed in the United States, eleven thousand people died, 5 billion dollars' worth of damages occurred, and thirty thousand cases of malaria, thirty thousand cases of cholera, and one thousand cases of dengue fever were reported.[47] This pattern has now been repeated a number of times. In Mozambique in 2000, for example, three cyclones and heavy rain over six weeks led to a fivefold increase in malaria.[48] Cholera outbreaks have been associated with warming and increased El Niño events since at least 1991, when an outbreak occurred in Peru and spread quickly to all its neighboring countries. Similarly, in India and Bangladesh, a dramatic new outbreak of cholera began in the port city of Madras in 1991, spread to other cities in southern India, then to Calcutta, and on to Bangladesh in 1992.[49]

Other debilitating mosquito and waterborne diseases also show strong correlations to climate change. These include dengue or hemorrhagic fever, which is carried by the *Aedes aegypti* mosquito, the same mosquito that transmits yellow fever, a disease Americans associate with Walter Reed, Teddy Roosevelt, and the Spanish-American War. Dengue or "break bone" fever, so-called because of extremely painful joint and bone pain, is, in the words of climate and health expert Dr. Jonathan Patz, "influenced by climate, including variability in temperature, moisture and solar radiation."[50] Other climate-related diseases showing increases include Ross River virus in Australia, plague in the American Southwest, and the spread of West Nile virus from the Mediterranean into the United States and across the nation.[51]

All told, the World Health Organization (WHO) estimates that warming and precipitation trends related to global climate change over the past thirty years have led to 150,000 deaths annually. But this figure only calculates deaths directly related to temperature changes such as heat waves, the spread of infectious disease, and famines related to drought. As Jonathan Patz has put it in describing the WHO estimates in which he participated: "the study made generally conservative assumptions about climate-health relationships. . . . And health impacts were included only if quantitative models were available." In sum, the figure of 150,000 deaths annually is a *very cautious* one. It is based only on direct heat effects that can be measured and for which there are studies. Indirect effects of climate such as disruptions from flooding or deaths from plain old air pollution—which gets worse and unhealthier with rising temperatures—are not included.[52]

Another major human health impact related to climate change is dirty, unhealthy air. It is now a worldwide phenomenon. All combustion of fossil fuel produces CO_2, leading to global warming. But the burning of gasoline for automobiles and other vehicles and of coal for electricity also gives off pollutants that cause or exacerbate heath conditions. These include asthma and other respiratory ailments, cardiovascular disease, and many cancers. Under various business-as-usual models of vehicle use and power production, these pollutants will continue to increase, causing harm to human health as well as pumping CO_2 into the atmosphere. Thus, the reductions in fossil fuel combustion needed to combat climate change would automatically lead to fewer pollutants and, therefore, to improvements in human health.

Worldwide, the health impact of air pollution is staggering. As early as 1997, just before Kyoto, Devra Davis and colleagues published a pioneering study in the respected British medical journal, *The Lancet.* It estimated that some seven hundred thousand preventable deaths annually, or about 8 million by the year 2020, are being caused by such pollution.[53] Again, the effects of air pollution are and will be most devastating in the rapidly urbanizing developing world. U.S. studies have already shown conclusively that ozone, an isotope of oxygen that forms when vehicle emissions are exposed to heat and sunlight, increases with temperature. As the globe heats up, the effects of air pollution on human health will only get worse. Indeed, according to a health study by Kim Knowlton and associates, of Columbia's Mailman School of Public Health, that uses careful modeling to reproduce regional effects, the greater New York area will see an increase of 4.5 percent in deaths by 2050 from ozone exposure alone. When population growth and other pollution factors are added in, the rates are substantially increased. The effects are seen as far away as the outer New York suburbs.[54] Whether in the developing world or in the United States, global warming will harm human health. But to get more people involved, in the next chapter, I will look at climate change, air pollution, and its other effects in places Americans tend to care most about—home, sweet home.

2

Home, Home on the Range: Climate Change in the United States

The adverse health effects of global climate change were noted by public health specialists well before and at Kyoto.[1] But the focus tended to be on places other than the United States. And the venues were usually journal articles, reports, and conferences where climate science, medicine, or policy was central—not popular public education. Given mind-numbing statistics and deaths, far from the average American's daily concerns (or news coverage), PSR decided to develop climate change studies accessible to the public, press, and policymakers. The reports featured material of interest to Americans where they lived, focusing on places and people they cared deeply about. Starting in 1998, PSR carried out studies and grassroots organizing campaigns in fifteen states. Called *Death by Degrees* (and later changed to *Degrees of Danger* out of sensitivity to the families who lost loved ones after the September 11 attacks), the reports were the first to systematically link climate change, human health effects, and local impacts.[2] *Death by Degrees* also implicitly drew on the political premise that even though the worst effects of global climate change were likely to occur outside the United States, progress on the global front would necessarily involve changing American attitudes.

The Power of Place

Like many people, my interest in climate change, temperatures, ecology, and human health perked up when I thought about areas close to my heart. When your hometown, or where your kids are growing up, or your favorite vacation spot is about to be ruined, it focuses your attention. Over the years, I learned lots from scientific conferences and papers on climate, air pollution, and other subjects. But I was sometimes bored or simply could not read the PowerPoint data slides that invariably accompany the standard scientific talk. It's why I still recall vividly the first time I was really jolted and felt personally affronted by global warming.

I was in Atlanta at an Environmental Protection Agency (EPA) conference. Colorful maps on PowerPoint slides showed browning and desertification, nighttime space photos captured lights visible across the globe, satellite infrared sensors revealed heat differences in the oceans. I could see visible storms from El Niño. But when the subject turned to sea-level rise and flooding, it was the picture of the Blackwater National Wildlife Refuge (NWR) on the Eastern Shore of Maryland that broke through my abstract, objectified interest.

Suddenly, I saw familiar outlines, water, peninsulas, land, dikes, marshes, actual dirt roads that I had traveled often and knew well. I watched in horror as I viewed projections of where Chesapeake Bay levels would be in the coming decades. I saw the land at Blackwater NWR slowly shrink, then disappear. I felt my stomach churn, my pulse race, and my anger rise along with the water. Blackwater was where my wife Caryn and I saw our very first American bald eagle and nest. It was a huge jumble of large sticks high up on a dead pine tree beside one of those dirt roads. It is where I first watched an eagle happily steal a fish from an osprey, land on shore, and pull it apart for lunch. Blackwater is where we also had offered our daughters a dime for each turtle they spied along the watery ditches that line the entrance road. We had regretted it as the dimes for turtles ticked off faster than a New York taxi meter. But now Blackwater would be inundated. Gone.

I was surprised, even shocked, at my own reaction. The eagles would plunder, perch, and presumably propagate elsewhere. These days, Caryn and I had little time for leisurely nature trips. But the loss of the familiar, of fond memories, of possibilities, haunted me. I was intellectually aware of phenomena like the connections between warming ocean temperatures and cholera outbreaks in Peru. But I have never been to Peru. I live in Maryland. I have walked and biked the C&O Canal and seen a phalanx of marching wild turkeys only seven miles from my suburban Bethesda house. I've witnessed foxes eating goslings, deer swimming, herons spearing fish or a frog, beavers chewing on maple saplings. I've watched gannets plunge like kamikazes into the winter ocean off Ocean City and seemingly prehistoric horseshoe crabs fling themselves ashore to lay eggs and die. And I love to see our national symbol, the bald eagle, circle, scream, and soar overhead in the sun—as if it and the United States were indeed invincible. All right here in Maryland.

It was at that moment when I saw and felt Blackwater wildlife refuge disappear, that I understood the power of place, of home, and the clichés—"all politics is local," "not in my backyard," "I Love New York," "Don't Mess with Texas." I resolved to try to bring global climate change home to people where they live. If I had some difficulty conjuring up outrage or sympathy for people in Peru—or the true meaning of 1 million to 2 million annual malaria deaths, mostly in Africa—I must not be alone. I cannot review all fifteen states in the *Death by Degrees* series here. But for this transplanted New Yorker, Maryland is a good place to start. Much of the data on climate change for Maryland comes from the EPA. But it has not been updated much during the Bush administration or during the Republican governorship of Robert Ehrlich that ended in 2007. But Maryland's current governor, Martin O'Malley, backed by environmentalists, has ordered new studies and a climate action plan.[3] Then, after Maryland, I'll head to New York and New England where I spent my youth. Then I'll go on to Texas, a critical state for our energy and political future. Finally, I will sum up some broad trends across the United States.

Climate and Maryland

In summer, Maryland can be hot and humid. It was originally a slave state, even though Lincoln maneuvered the state legislature into declaring for the Union. Frederick Douglass was a slave and taught himself to read on the Eastern Shore. Harriet Tubman led slaves to freedom from here, and Harriet Beecher Stowe's Uncle Tom (historically the Rev. Josiah Henson), who escaped across the frozen Ohio River with Little Liza, lived in a slave cabin and toiled in tobacco fields not far from my home in Bethesda.[4] So you know Maryland can be hot. Field hand hot. Slave hot. And those temperatures are getting worse.

In the past couple of hundred years, roughly since the invention of Eli Whitney's cotton gin and the expansion of slavery in the United States, the rate of increase in global temperature has been accelerating. In just the past century, average global temperatures have increased by more than 1°F. But in Maryland, over the same time period, the average temperature in College Park, home of the University of Maryland, has risen by 2.4°F. Precipitation has increased by 10 percent in many parts of the state. This trend toward warmer and wetter weather is predicted to continue. Over the course of this century, projections from the IPCC and the United Kingdom Hadley Centre Model, show temperatures in Maryland could increase, in mid-range estimates, by 3°F in spring and 4°F in other seasons.[5]

I had suspected that summers were getting worse in Maryland, but I couldn't be sure. I don't keep weather records. I mainly sweat and complain. And so, I felt somewhat personally vindicated by a study we did at PSR with the environmental group, Ozone Action. We used data gathered by NOAA from 171 weather stations from 1948 to 1998. Then we extended the NOAA data and methodology until 2001. There it was. The records and graphs showed it unmistakably. The number of heat stress days and nights across the country had doubled in the past fifty years. The number of four-day heat waves had nearly tripled.[6]

Those charts not only signaled hotter summers. They were a clear sign of danger to human health. High humidity is dangerous because it inhibits the body's ability to cool itself through evaporative heat loss and perspiration. This leads to stress, and in some cases, death. In fact, annually, heat waves are more deadly than any other natural disaster in the United States. According to the National Center for Health Statistics (NCHS), there have been, on average over the past thirty years, 371 heat deaths per year. That's more than from hurricanes, tornadoes, and flooding combined.[7] During extreme heat, excess mortality may be even higher. In 1995, a five-day heat wave in Chicago was linked to more than seven hundred deaths. Even these numbers are probably underreported, as many heat deaths are not registered as such on death certificates.[8] A study by Kalkstein and Greene at the University of Delaware Center for Climatic Research examined mortality and weather data for U.S. cities, including Baltimore. They found significant increases in deaths per day during heat waves, especially among the elderly. Excess summer heat deaths just in Baltimore are projected to roughly double from 84 to 164 by 2050.[9]

Closely related to rising temperature is increased air pollution. In a rapidly urbanizing Maryland, it will have seriously adverse effects on health. Air pollution, with the bulk coming from vehicles and power plants, causes some fifty thousand deaths annually in the United States. Climate change adds to the problem, as consumers draw more power for air-conditioning in peak summer periods. Because CO_2 is one of the gases given off, as citizens try to adapt, the greenhouse effect is exacerbated. Maryland ranked seventeenth nationally in carbon emission from power plants in a 1997 study, releasing 49.7 million tons of carbon. The largest CO_2 emitters were the Herbert A. Wagner plant operated by Baltimore Gas & Electric with 12.4 million tons, BG&E's Brandon Shores with 10 million tons, and BG&E's Gould Street Plant with 9.2 million tons.[10]

Power plants also emit nitrous oxides, or NOx, a greenhouse gas that, according to the EPA, accounted for 7 percent of U.S. greenhouse gas emissions in 2005.[11] When combined with high temperatures and volatile organic compounds (VOCs), NOx also produces ozone or smog. An estimated 117 million Americans live in areas where the air is unsafe to breathe owing to ozone. In 1999, the ozone health standard adopted by EPA in 1997 was exceeded 7,200 times.[12] According to a study of Maryland power plants by the Clean Air Network, Maryland also ranked seventeenth in NOx emissions with a total of 103,322 tons. According to a 2007 report, the worst emitters in the state now are the Chalk Point plant in Prince George's County with 14,043 tons, the Morgantown plant of Potomac Electric Power with 13,759 tons, BG&E's Brandon Shores with 11,893 tons, and CP Crane of BG&E in Baltimore with 7,705 tons.[13]

Exposure to elevated ozone levels can cause severe coughing, shortness of breath, lung and eye irritation, and greater susceptibility to respiratory illnesses such as bronchitis and pneumonia. Even moderately exercising healthy adults can experience a 15 to more than 20 percent reduction in lung function from exposure to low levels of ozone.[14] It is a special concern for asthmatics. Numerous studies have shown it causes more attacks, increases the need for medication and medical treatment, and results in more hospital admissions and emergency room visits.[15] One study of twenty-five hospitals found that high ozone levels were associated with a 21 percent increase in ER visits for persons sixty-four years and older.[16] An EPA study of Baltimore, for example, showed an increase in ozone alert days with the air quality index (AQI) over 100 on average 37.4 days per year from 1990 to 1999.[17] The result? Maryland experienced 180,000 ozone-related asthma attacks, 1,300 hospital visits, 3,900 visits to ERs for ozone-related respiratory illnesses, and 410 hospital admissions for cardiovascular events.[18] Baltimore alone suffered 86,000 asthma attacks, 630 hospital visits, 1,890 ER visits, and 200 cardiovascular hospital admissions.[19] An independent study by the Maryland Nurses Association looked at data from only six major coal-fired utilities. It found roughly one hundred premature deaths and four thousand asthma cases each year attributable to coal burning for electricity.[20]

In addition to heat and air pollution, Maryland is also especially vulnerable to sea level rise. The National Academy of Sciences predicts sea level rise of about 2.3 feet along the Atlantic Coast by 2100, while an EPA mid- to high-level scenario shows a rise of about 4.5 feet.[21] Maryland, famed for the beauty, productivity, and recreation provided by the Chesapeake Bay, has 3,100 miles of tidally influenced shoreline. The bay covers 4,402 square miles, and its watershed drains a region of 64,016 square miles. Maryland brings to mind sailboats, sandy shores, and crab suppers spiced with Old Bay. The Chesapeake is one of the best studied and most historic areas among those first settled by Europeans. It has been undergoing sea level rise for some ten thousand years since the glaciers retreated. But over the past six thousand years, that rise has been only about 6 inches per century. Increases in sea level are clear from early colonial maps and from the submergence of Sharps Island, which was mapped by John Smith in 1608. Today Sharps is covered by 3 to 4 feet of water with only a ruined lighthouse built in 1882 still visible.[22] Current sea level rise in Maryland is about 1.3 feet per century. With further global warming that rate is expected to double within just the next twenty-five years. By 2100, the EPA mid-range estimates show a rise of 4.5 feet. That would threaten developed beachfronts, such as Ocean City; barrier islands, such as Assateague Island and its wild ponies; and wildlife and birding areas on the Eastern Shore with its extensive freshwater and salt marshes, forested wetlands, and open water.[23] In Baltimore, our major port, sea level is already rising at 7 inches per century. It is likely, given climate change, to rise at least another 19 inches by 2100.[24] All told, about 600 square miles of Maryland land is vulnerable. Over just the next forty years, sea level rise in Ocean City could double from an expected 15-inch increase, causing beach erosion of some 275 feet.[25] The cost of additional sand replenishment to protect Maryland's beaches from a 20-inch sea level rise by 2100 has been estimated at up to $200 million.[26] As the ocean moves in and wetlands disappear, the ground and surface water upon which human health depends is also compromised through increased salinity. Salinity has also led to declining oyster harvests in the Chesapeake Bay, further undermining food sources and the economic livelihood of the area.[27]

With reductions in wetlands and forests, problems of water contamination and loss of natural filtration are especially pressing. Maryland is a state undergoing rapid urban sprawl.[28] It is estimated that in twenty-five years—assuming growth continues in the Chesapeake watershed at the same rate as the 1990s—more than 3,500 square miles of forests, wetlands, and farms will become urbanized. That's equivalent to fifty Washington, D.C.'s. Such sprawl uses additional energy for homes, businesses, and driving. In the Chesapeake watershed area alone, vehicle miles traveled (VMTs) have increased by 105 percent while population has grown by only 26 percent.[29] All this creates more greenhouse emissions and increases the amount of paved, impervious ground. That exacerbates erosion, runoff, and water pollution.

Global climate change, with its increase in extreme weather events and heavy rains, aggravates the situation. It will further compromise water quality and lead to waterborne diseases. A study by Jonathan Patz and Joan Rose reported that between 1971 and 1994, 20–40 percent of drinking water–related outbreaks of disease in the United States were associated with extreme rainfall.[30] Downpours caused by climate change bring on additional human hazards like floods, which already cause an average of 146 deaths per year nationwide. Most at risk are those who live in floodplains or areas where forests have been leveled for development.[31] That description fits Maryland. According to a 2000 survey of weather-related losses and government spending, Florida, Texas, and California topped the list. But Maryland, which does not have the huge population or devastating hurricanes and earthquakes of those states, still spent nearly a billion dollars on storm losses in the 1990s.[32]

In addition to the drownings and injuries that come with flooding and high water (about seven people a year drown in the Potomac River in a single spot I pass on my bike rides along the C&O Canal), flood waters may contain human and animal fecal material from overflowing sewage systems and flooded croplands. Besides *E. coli,* contaminated waters can cause disease from a variety of bacteria (e.g., *Salmonella* and *Shigella*), viruses (e.g., rotavirus), and protozoa (e.g. *Giardia lamblia,* amoebas, *cryptosporidium*).[33] *Cryptosporidium parvum* is a waterborne parasite that produces life-threatening zoonosis in people with compromised immune systems, such as cancer patients or the elderly. It caused 54 deaths and 403,000 illnesses in a well-publicized case in Milwaukee. But *Cryptosporidium* oocysts have been shown to increase with rainfall in places like the Delaware River.[34]

Maryland, whose watershed includes significant cattle areas upstream in Pennsylvania and within the state, is vulnerable. A study by Graczyk and others, for example, found *Cryptosporidium* in 60 percent of all manure samples on cattle farms in Lancaster, Pennsylvania.[35] In such bucolic areas, typical of the Chesapeake Bay watershed, cattle are allowed to roam freely in the many small streams and rivulets. As temperatures in Maryland waters increase, estuaries tend to hold less oxygen and to form stratified layers with lighter freshwater lying on top of layers of dense, heavy saltwater. The bottom layer is then cut off from almost all oxygen, a condition that NOAA recognizes as a precursor to the development of *Pfiesteria.* That's an organism that can change back and forth from plant to animal and infect fish, putting humans, especially watermen in Maryland estuaries, at risk. Dr. JoAnn Burkholder, for example, investigated fish kills in North Carolina. She and a colleague became dizzy, disoriented, and unable to speak or walk after studying samples.[36]

Finally, the Eastern Shore of Maryland poses particular risks to human health, especially in the face of sea level rise, flooding, and warmer temperatures. It is home to the factory-style production of chickens owned or contracted by Perdue and other companies. These vast confining areas and slaughterhouses turn out large amounts of manure and other wastes—a growing source of contamination.[37]

Communicating local climate effects on human health and well-being and drawing on the power of place can be tricky, however. In teaching climate change in the Master's Program in Global Environmental Politics at American University, I eagerly showed an award-winning film on climate and Maryland called *We Are All Smith Islanders.*[38] I assumed, rightly, that a number of my students would be from Maryland or have visited the Eastern Shore. What I had not counted on was that the film's primary vehicle for localizing and humanizing the subject was interviews with Chesapeake Bay watermen. They are a fascinating, if dwindling, group. They make their living off the increasingly difficult job of fishing for crabs and oysters in the endless inlets and estuaries of the bay. To be a successful waterman requires careful observation and understanding of the many moods of the Chesapeake, its weather, and terrain. To a person, the men and their families on screen testified that temperatures had increased and that water levels had risen and were threatening Smith Island. Climate was indeed a problem.

But the problem for my class—all dedicated environmentalists and a number of them active in agencies like EPA—was that they were urban to the core, young, and used to much faster entertainment and education.

Most could not understand the distinctive dialect of the watermen who had lived in relative isolation for generations on Smith Island. They were not unsympathetic to the filmmakers, the watermen, or the wildlife. They simply were not moved emotionally in the ways I had anticipated as they watched Smith Island being submerged and ruined before their eyes. Lest you think American University students imaginative and moral cretins, they responded quite strongly when I described as graphically and disgustingly as I could, during my own PowerPoint show on climate in Maryland, the effects of increased sea levels and heavy storm rains in the Baltimore Harbor. It is a favorite tourist site with a mall, an excellent aquarium featuring sharks and beluga whales, waterside restaurants, boat rides, breweries. Most of my students had been there.

In Baltimore, as in many parts of the country, drinking water supplies and storm sewage treatment plants are connected. During heavy rains that overflow storm drainage, drinking water can become contaminated. In one year, according to Maryland State Representative Dan Morhaim, a physician who serves the Eleventh District and sits on the environment committee, there were twelve large sewage overflows of 100,000 gallons or more in Baltimore. There were also seventy-five medium spills of between 10,000 and 100,000 gallons.[39] Overall, millions of gallons of raw sewage have been dumped into Chesapeake Bay estuaries. In one of the worst Maryland overflows in 1999, 24 million gallons of raw sewage spilled from a pumping station in Baltimore's Inner Harbor when rain flooded the facility. The result, aside from that familiar sewage stench, is a rapid rise in *E. coli* counts that causes diarrhea and other gastrointestinal problems.[40] My students had seen and sniffed such malodorous results of "extreme weather events" and "increased precipitation" both in Baltimore and along the Potomac in Georgetown. After a storm, you can smell effluent near the quaint clapboard crew boathouses that sit at the bottom of the popular Crescent biking and jogging trail. It's not far from Sequoia, one of the most popular waterfront hangouts. That and Inner Harbor Baltimore really got their attention. But they were concerned that climate problems are not merely local, that not enough Americans care about marshes. The problems, they said, are regional and national, too.

New York, New York

That's why many of my students were engaged by the greater New York–New Jersey metropolitan region where I grew up. They knew it pretty well—some as residents, others from hitting the beaches, touring the sights and entertainment of Manhattan, or visiting friends. Authors Jim Motavalli and Sherry Barnes describe the 1992 nor'easter that hit New York City. It caused water surges 8.5 feet above normal, bringing the city's transportation infrastructure to a halt. Four million subway riders were stranded. Flights were grounded at LaGuardia Airport, which sits by the water only 7 feet above sea level. FDR Drive, running down the east side of Manhattan, flooded up to 4.5 feet. The scenario was repeated in 1999, during the summer, when a heat wave knocked out power for two hundred thousand people in Manhattan. Thirty-three people died. Rainfall in New York was 8 inches below normal, and drought conditions prevailed until heavy rains again hit the city in late August, flooding both the FDR and West Side Drives. Subway tracks were submerged in 5 feet of water. Hurricane Floyd followed soon after with $1 billion in damages.[41] By the end of this century, under mid-range IPCC scenarios, Janine Bloomfield, who wrote the Environmental Defense report "Hot Nights in the City," points out that New York will have as many 90° days as Miami does now. A large part of lower Manhattan, the financial center of the nation, will be subject to frequent flooding. And, of course, for birders like me, one more key preserve, like Blackwater, will be gone. The Jamaica Bay refuge where I have observed thousands of migrating shorebirds is what is left of once extensive marshlands. They have been shrinking by 3 percent per year as a result of both climate change and sea level rise. I first visited Jamaica Bay as a boy when Kennedy Airport, then known as Idlewild, was new. It consisted of only World War II Quonset huts and runways. Now, unless action is taken, the birds, the marshes, and two of the busiest airports in the nation will be inundated.[42]

Again, the health effects of climate change on New Yorkers are sadly impressive. With temperatures this century slated to rise by at least 4°F, air pollution and its related diseases will be exacerbated. The American Lung Association reported that given ozone levels at the end of the twentieth century, 2.3 million New Yorkers were at risk of respiratory problems. And it is from New York that one of the most serious and spreading insect-borne diseases, West Nile Virus has come. First identified in Uganda in 1937, it is an animal-related disease ("zoonosis") that can spill over to humans. West

Nile was unknown in the Americas until it appeared in Queens, New York, in the summer of 1999, leading to seven deaths and numerous cases of nervous system disorders. It is uncertain how the virus that is spread by mosquitoes first arrived. But, once introduced, warmer temperatures nationwide allowed the disease to spread from mosquitoes to hosts, including birds and small animals, and then to humans. In the United States, West Nile virus has now been reported in every state except Oregon, Alaska, and Hawaii. Carried by the *Culex pipiens* mosquito, West Nile is associated with droughts. It flourishes in storm-catch basins where nutrient-rich water remains after a drought. The shrinking availability of water also concentrates the birds. Near water, they are often carriers of the disease because they offer more frequent blood meals to mosquitoes.

In the original New York City outbreak, sixty-two New Yorkers developed nervous system disease: encephalitis (brain inflammation) and meningoencephalitis (inflammation of the brain and surrounding membrane), with seven deaths and persistent neurological impairment for most of the remaining. Since that time, up to the end of 2005, there has been a significant increase in West Nile among Americans nationwide: 4,161 cases and 284 deaths in 2002; 9,856 cases and 262 deaths in 2003; 7,470 cases and 88 deaths in 2004. In 2005 and 2006, reported cases declined somewhat. But fatalities continued to increase to 119 and 177.[43] As bad as these effects are for New York (and the rest of the country as diseases spread or other ripple effects are felt), they pale compared with more extreme scenarios and possibilities such as the huge sea level rise that would accompany the melting of the Antarctica and Greenland ice covers. As shown in *An Inconvenient Truth,* such a catastrophe would flood much of metropolitan New York. Some 20 million people and a significant portion of the nation's financial, media, and cultural institutions are there.[44]

New England

But it is not just beaches and cities that are at risk. Like many East Coast natives, I have vacationed throughout New England. I was surprised, irritated, and again determined to do something when I discovered that even the cool, green granite mountains of New Hampshire are seriously affected.

PSR had published and organized in New Hampshire around our very first *Death by Degrees* report during the presidential primary season of 2000. We released its findings, along with groups like the Sierra Club and Ozone Action, at the state Capitol Building in Concord. We aimed at the media, the general public, state legislators, and the anti-Kyoto efforts of the Chamber of Commerce. In New Hampshire—that summer Shangri-la where vacationers can cool off; inhale fresh, piney air; and take in woods, mountain views, and lakes—one of the worst effects of global warming, surprisingly, is respiratory disease. It is brought on and exacerbated by increasing ozone levels. Long ago, as a boy on family summer vacation in the 1950s, I had stood atop Mt. Washington, one of the highest mountains in the East. I remember marveling at the windy, wintry air whipping at my mom and sister's scarves and views as far off as Montreal. A generation later, I had repeated this ritual with my own kids for the views and the iconic bumper sticker ("This Car Climbed Mt. Washington"). The wintry winds and views seemed as wonderful as I had remembered. But when I released the PSR report in front of the State Capitol with Dr. Jim Herriod, a physician and Republican state senator, and Representative Barbara French, a Democratic state representative and nurse, I was appalled. Ozone levels regularly reported atop Mt. Washington were now failing to meet EPA standards! As with most things related to climate change and pollution, it is scientific measurement and trends, not just anecdotes and memories that count. Air pollution in New Hampshire had been increasing steadily from cars, power plants, and drift from other states. And it is made worse by rising temperatures.[45] The overall prevalence of asthma rose 72 percent from 1982 to 1994, while the death rate increased by 78 percent during a similar period. Asthma accounts for one-sixth of pediatric emergency room visits in the United States, and even in New Hampshire, a small, green state, we found in 1998 that asthma accounted for 6.6 percent of pediatric emergency cases.[46]

Our New Hampshire research also showed that extreme weather events were already causing damage, with a prolonged drought and water shortages in 1997, a huge ice storm in 1998, and increased summer heat waves and eleven more frost-free days than only two decades before. New Hampshire also suffered from other climate effects, including red tides that create toxic shellfish and sea level rise that according to current projections would swamp the only part of New Hampshire that touches the ocean—historic and carefully restored Portsmouth.[47]

PSR also produced separate reports in the Northeast on Maine, New York, and Pennsylvania. I know and love each of these places. But like my students, I'm interested in regions, too. Having grown up on Long Island, I still long to walk on Jones Beach. But in my youth, I realized that the boundaries between New York, New Jersey, and Connecticut did not mean much for urban or environmental problems. And after the interstate highways were built, New Hampshire and Vermont, Massachusetts and Maine did not seem far away either. There are some real similarities and vulnerabilities throughout the entire Northeast. Since the Bush administration stopped producing much new, useful information on climate change in the United States, it is only in 2007 that we got a fresh look at what is likely to happen in nine states in the Northeast. The Union of Concerned Scientists (UCS), an independent nonprofit group, assembled an expert panel of fifty scientists and economists to review recent studies and produce a peer-reviewed report. Peter Frumhoff, director of science and policy at the University of Southern California, and a member of the most recent 2007 IPCC study on the global effects of climate change, says: "the very character of the Northeast is at stake." UCS found that New York City could be hit by the extreme flooding already described in every decade instead of every hundred years. Even worse, based on the latest studies, Atlantic City, New Jersey, and Boston could be hit by such floods every two years. That includes the possibility, says Frumhoff, of storms similar to the 1938 hurricane that killed hundreds and swept away thousands of buildings. By 2100, without strong action, average temperatures could be 10° hotter than now. Cities throughout the Northeast could also suffer through twenty-five days a year of temperatures above 100°F, while shade trees like maples and hemlocks could die from parasites and the warmer weather. A warmer climate would also destroy regional industries such as skiing and lobstering.[48]

Don't Mess with Texas

As I travel the country talking about climate change, the area that I have visited over the years that fascinated me most has been Texas. It is huge. It is almost an entire region unto itself. It ranges from coastal beaches, islands, and estuaries on the Gulf Coast to pleasant highlands to the arid, sunny Southwest. Like many older baby boomers, I grew up with myths and images of Texas based on Hollywood lore. Cattle drives, John Wayne, tumblin' tumbleweed, and vast open stretches—perhaps dotted by a few oil wells—were what we imagined. I only began to visit regularly in the 1990s when Dr. Howard Frumkin, now head of the Centers for Disease Control and Prevention (CDC)'s combined National Center for Environmental Health/Agency for Toxic Substances Disease Registry (NCEH/ATSDR), and I set up a PSR United States–Mexico Border Project in El Paso/Ciudad Juárez. By then, El Paso was overwhelmingly Latino. It had severe air and water pollution problems related to gas refineries, brick-making kilns, pesticide run off, and chemical exposures from small assembly factories called *maquilladoras.* They dot the Mexican side of the border. But those who suffer adverse health effects from them often come to emergency rooms and hospitals on the U.S. side. Asthma rates on both sides of the border are particularly high, and the PSR project, along with Environmental Defense, focused on air quality for the entire Paso del Norte air quality management district.[49] I recalled all this and the heat, temperature inversions, and air pollution of Texas when PSR began its *Death by Degrees* series. Texas, our second-largest state in land and population, continues to grow in political importance. It had just given the nation its third president in my lifetime. It seemed a good place to look.

In the Lone Star State, gone are the days of the Alamo, the Mexican-American War, or the Wild West. Now heavily industrialized and urbanized, with 80 percent of Texans living in cities, Texas generates 99 percent of its electricity from coal, oil, natural gas, and nuclear power. It is the nation's largest producer of CO_2 and unhealthy air emissions. If Texas were an independent nation, it would rank seventh in the world in CO_2 production. It ranks the same as the United Kingdom, which has triple the population. Since 1995, Texas has emitted close to twice as much CO_2 as any other state in the union. It accounts for more than 10 percent of the U.S. total. Each Texan is responsible for the production of 10.4 tons of carbon (38.5 tons of CO_2)—nearly twice the national average. Compared to most states, Texas has a much greater share of greenhouse emissions from industry. That's because it has twenty-six oil refineries. It also has a high percentage of electricity produced from coal. Without changes in state, national, and international policy on carbon emissions and renewable standards, Texas is currently projected in most seasons to have temperature increases of 4°F with a possible range up to 9°F. Rainfall could drop up to 30 percent in winter, while increasing by 10 percent in other seasons, especially summer, when it could increase by 30 percent. Such uneven precipitation will lead to increased drought, flash floods, and storm damage.[50]

Texas is remarkably diverse—in its population of Anglos, Chicanos, African Americans, and others—and in its weather extremes. They range from drought that parches the plains of the Panhandle and west and southwest Texas, hurricanes that wreak havoc along the Gulf Coast, and summer heat waves that broil Houston-Galveston, Dallas–Fort Worth, and other large cities. Ask any American about smog, and the response will be "Los Angeles." But for years Houston, the fourth-largest city in the country, has been vying with L.A. for the title of most polluted city. It has already beaten the cinema capital several times. In the last two decades of the twentieth century alone, Texas also suffered more than sixteen weather disasters costing over $1 billion each. And Texas, according to the 2003 PSR report, had led the nation in heat-related deaths three out of the previous seven years. In 1998 alone, a record-breaking heat and ozone year in Texas, chronic respiratory diseases related to pollution and temperature resulted in $435 million in direct and $328 million in indirect medical expenditures. That year, Texas ranked seventh in the nation in pediatric asthma, while more than 2.5 million Texan kids (2,528,719) lived in counties that exceeded the EPA's eight-hour ozone standard.[51] Texas also has the highest levels in the nation of toxic mercury emissions. These are a result of burning coal in electric power plants and result in problems in the neurological development of children, seizures, and even death.[52]

Texas, like much of the Southwest, also faces severe problems with water. The state is already depleting its water resources at an alarming rate. Many of its sources of groundwater are overdrawn, especially along the Rio Grande. It now trickles through El Paso in a small, pathetic, dark stream amid concrete channels. El Paso's aquifer may be depleted as early as 2030. And, with Texas population expected to double to about 40 million by 2050, without strict state water plans and conservation policies, municipal and industrial water use will surely grow and strain supplies. Climate change will make things worse. Predictions are for a decrease of rain in winter and an increase of 20 percent in other seasons, mostly in very wet days. These result in rapid runoff and flash floods, yet negligible increases in state water supplies. Water quality is a serious problem in Texas, too. Its floods and droughts, as we saw in Maryland, can lead to water contaminated with microorganisms. In the border areas, shigellosis and hepatitis A rates are already triple the national average.[53]

Nationwide Effects

PSR's *Death by Degrees* kick-off presentations in each state were followed by educational events at medical schools, hospitals, universities, and civic gatherings nationwide. Although they clearly show state and regional variations, several main themes emerge from various reports on climate change across the United States. These include serious effects from sea level rise, flooding, wetland destruction, and the spread of waterborne and vector-borne diseases in coastal areas. This is especially true along the East Coast with its strings of barrier islands and beaches from New England through Florida. The latter has become a sort of poster child for sea level rise, because in scenarios of the future, large portions of the state, starting with the Keys and the Everglades, will be lost. Yet even as rain increases and shorelines rise, drought will also be a major effect of climate change. As 2007 ended, I was in residence for a week at Berry College in Rome, Georgia. I witnessed the severe drought in northwest Georgia that disrupted agriculture and led to legal battles over dwindling water supplies in Georgia, Alabama, and Florida. The PSR *Death by Degrees* report had quoted the state climatologist, David Stooksbury, about climate change and the Georgia drought of 1998–2000. The study warned of severe future ones if action was not taken. The warnings went essentially unheeded. But in 2007 and early 2008, when Stooksbury was forced to repeat the same message, Berry College and its president, Steven Briggs, who had just signed a campus sustainability pledge, as well as the entire local area connected its parched land with global warming.[54] Though not as severe as in Georgia, drought gripped typically more lush areas throughout the East Coast as well. The midwestern states also saw serious drought rivaling that of the 1930s Dust Bowl, while wildfires were an increasing problem in the West. California, another state that is practically a region, suffered extensive fires in late 2007. The Golden State also has notorious air pollution. Its extent and health effects are well described in Terry Tamminen's *Lives per Gallon: The True Cost of Our Oil Addiction.*[55] But, along with states in the Pacific Northwest that rely on snow pack or glaciers for freshwater supply, California's supplies and quality of water are also at risk. With rising temperatures, waterborne diseases will also increase, and agriculture will be threatened. And, of course, in every region, as temperatures rise, urban air pollution and its pervasive health risks will increase and spread.

As I summarize somewhat tongue-in-cheek for students, put simply, global warming is just plain bad for your health. From a public health and policy standpoint, we have a serious, documented, physically and biologically plausible, and increasingly dangerous situation on our hands. But if, in broad terms, we have understood much of this since the Carter administration, why has it taken so long to prove it and to get action? What forces are we up against, and what are some reasonable strategies and targets? There is no simple answer or drug injection, if you will, to prevent climate change. But a key obstacle has been the power and political influence of the industries that extract and produce fossil fuels—oil, coal, and gas. So are related businesses, like automobiles, utilities, and chemicals that rely on them. I mentioned that scientific research and scientific policymaking are never as objective and pure as many people believe. In fact, the machinations of the carbon lobby are a case study of how industries often affect basic science and policy. That's why in the next chapter, I will look at why it is that you and I still meet people innocently repeating long-disproved assertions about the scientific uncertainties, costs, and difficulties of moving away from fossil fuels, reducing CO_2 emissions, and creating a healthy, sustainable energy future.

3

The Power of the Carbon Lobby

Some American public health problems—like the dangers of lead in gasoline or smoking cigarettes were well known for many decades before any effective public action was taken. The reason? Powerful corporations like the Ethyl Corporation in the case of lead, or R. J. Reynolds and others with tobacco, waged lengthy lobbying and public relations campaigns to prevent their products from being regulated or restricted. These corporate giants were able to advertise widely, to subsidize scientists, and to prowl Capitol Hill—all while dispensing generous political contributions. Not surprisingly, doctors, public health officials, and consumer advocates had difficulty being heard. They had to wage lengthy, determined campaigns to finally convince the public and policymakers of the unvarnished truth. Global climate change presents similar challenges. It, too, will be remembered as a textbook example of corporate greed and disinformation that for far too long outweighed the public good.[1]

Oil and National Security

The powerful influence of oil in U.S. policy stretches back across administrations of both parties to that of Franklin Delano Roosevelt. While hosting a PSR seminar for congressional and environmental staffers on oil, climate change, and national security, I recall my combination of shock and mild amusement at the photo put up on screen by Michael Klare, professor of global security studies at Hampshire College. He is an expert on natural resources and war. Klare, whose noticeable New York accent gives special bite to points he wants to make, gestured toward the wall: "This is where it begins." There was FDR aboard the American naval cruiser USS *Quincy,* anchored at the southern end of the Suez Canal on February 14, 1945, toward the end of World War II. FDR is engaged in lively conversation with

Saudi King Ibn Saud, in robes now widely familiar from various news stories and documentaries about the influence of the rich oil kingdom. Roosevelt had flown directly to Egypt from the Yalta Conference with Churchill and Stalin. After discussion of a number of Middle East subjects, according to Klare, Roosevelt and Ibn Saud "forged a tacit alliance—one which obliged the United States to protect Saudi sovereignty and independence in return for a Saudi pledge to uphold the American firms' dominance of the oil fields."[2]

Oil had emerged as one of the crucial factors in World War II, not only in Hitler's attempts to control it in eastern Europe, Russia, and the Middle East, but in the Allied victory, which depended on huge fleets of ships, airplanes, tanks, and other vehicles that gave the margin of success. Between Pearl Harbor and the defeat of Japan, the United States supplied 6 billion of the 7 billion barrels of oil that Allied forces consumed. Foreshadowing contemporary discussions of oil and national security, U.S. officials monitoring oil supplies were aware and alarmed that domestic oil production was peaking and would not be sufficient to sustain the country's needs as the superpower to emerge after the war. Oil had become a strategic need. Concerns that U.S. known reserves of 20 billion barrels in 1941 would run out in the postwar period, imperiling American prosperity and power, led that year to the first systematic assessment of the strategic need for overseas oil. In it, William Ferris of the State Department concluded that the United States should follow "a more and more aggressive foreign oil policy aimed at assuring access to petroleum overseas." This approach was codified in April 1944 with the issuance of the State Department's *Petroleum Policy of the United States.* It recommended that the United States should pursue "the orderly and substantial expansion of production in eastern hemisphere sources of supply, principally the Middle East."[3]

Once committed to promoting stability and access to oil in the Middle East, the United States has consistently adhered to this policy, beginning with the Truman Doctrine and the first Cold War clash with the Soviet Union in 1946 over the oil fields of Iran. The United States not only declared in 1943 that "the defense of Saudi Arabia is vital to the defense of the United States." It also relied heavily on the British as the regional authority to maintain oil supplies and maintain stability. But when the British announced in 1968 that they were pulling out of the Middle East, the United States developed the "surrogate strategy." It drew on the Nixon Doctrine and relied on its then powerful allies in the area—Iran under the Shah, and Saudi Arabia—to protect U.S. interests. Billions of dollars of sophisticated weaponry was sent to Iran and Saudi Arabia, including advanced fighters, tanks, and attack helicopters. In Saudi Arabia, much of the equipment went to the Saudi Arabian National Guard, or SANG, whose new headquarters was built by the U.S. Army Corps of Engineers. It became an early target of Osama bin Laden. Along with the large quantities of weapons came American military advisers or troops who, by 1977, numbered over ten thousand in the two countries.[4]

When in 1979 Islamic militants loyal to the Ayatollah Ruhollah Khomeini overthrew the Shah of Iran, President Jimmy Carter determined that the United States would need to take on responsibility for protecting access to Persian Gulf oil. Thus was born the Carter Doctrine, which the president announced in his final State of the Union address on January 23, 1980. It is still reigning U.S. policy today: "An attempt by any outside force to gain control of the Persian Gulf region will be regarded as an assault on the vital interests of the United States of America, and such an assault will be repelled by any means necessary, including military force."[5] Under the Carter Doctrine, the United States developed a Middle East deployment force that ultimately became the U.S. Central Command, procured long-range cargo planes, and negotiated basing rights in Oman, Kenya, Somalia, and on Diego Garcia in the Indian Ocean. When President Reagan took office, he then vastly increased military support for Saudi Arabia, including advanced AWACs aircraft over the objections of Congress. But he got the Saudis in return to financially support the CIA's secret efforts to overthrow the Soviet-backed regimes in Afghanistan, Nicaragua, and elsewhere. Among those who benefited from this Saudi and CIA support was, of course, Osama bin Laden. He helped recruit religious zealots to fight in Afghanistan, thus laying the foundations for the Taliban and al-Qaida. The United States then also provided support to Saddam Hussein and escorted reflagged Kuwaiti tankers through the Persian Gulf during the Iran-Iraq War, finally causing the Iranians to give up the fight. When Saddam later seized Kuwaiti oil fields and control over 25 percent of the world's oil reserves as a way to relieve debt Iraq had accumulated during the war, President George H. W. Bush quickly moved U.S. troops into Saudi Arabia to defend it just four days after the Iraqi invasion of Kuwait. In the earliest pronouncements about the Persian Gulf War, both President Bush and Secretary of Defense Dick Cheney spoke of the role of oil in the war. President Bush specifically noted that the United States now "imports nearly half the oil it consumes and could face a major threat to its economic independence." Cheney, in turn, said to the Senate Armed Services Committee that if Saddam Hussein were able to maintain control of Kuwait, he would "be able to dictate the future of world energy policy, and that [would give] him a stranglehold on our economy."[6]

By the time of the Iraq War and the U.S. decision to invade without evidence of weapons of mass destruction, it was clear that oil played a significant role in the choice. The most candid remark along these lines came again from Dick Cheney in an August 2002 speech to the Veterans of Foreign Wars. He said: "Armed with an arsenal of these weapons (WMD) and a seat atop 10 percent of the world's oil reserves, Saddam Hussein could then be expected to seek domination of the entire Middle East, take control of a great portion of the world's energy supplies, directly threaten America's friends throughout the region, and subject the United States or any other nation to nuclear blackmail."[7]

The United States invaded Iraq in March 2003 as part of a wider "war on terror." But U.S. forces moved quickly to protect Iraqi oil infrastructure, including oil terminals north of Kuwait and the oilfields near Basra. Protecting other assets or gaining control of key areas on the ground came second. The war on terror is indeed linked to radical Islamic fundamentalism (although attempts to link Saddam Hussein to al-Qaida failed and were never retracted by the administration). But the seeds of attacks on U.S. interests and allies in the region and elsewhere can be traced back to a fateful U.S. dependence on Middle East oil and the policies to defend it that stretch all the way back to a handshake aboard the USS *Quincy.* The influence of oil and the oil lobby in recent years has purposely set out to confuse the American public about the science and importance of global climate change and its links to fossil fuels like oil. At the same time, the relentless focus on a military approach to national security and protecting U.S. interests has frequently driven concern for global warming as a serious long-term threat to U.S. security out of visible media and political concern. Only with the slow rise of opposition in the United States to the Iraq War and the realization that the Bush administration had lied about the evidence proffered to gain support for it, driven the nation back into debt, and caused oil supplies to be threatened and prices to skyrocket, would there be sufficient news and political space to once again raise the specter of global climate change. Until now, the effort led by environmentalists and scientists to alert the world to the dangers of climate change and oil dependence had been pretty much a David versus Goliath affair.

THE CARBON LOBBY AND SCIENCE

As scientific evidence and environmental concern about climate change grew throughout the 1980s, so did the response of the industries—oil, coal, and gas—that produced the CO_2 emissions acting to warm the planet. A decades-long battle over the facts of global warming, far from the mainstream evening news, was about to begin. It would feature a small number of scientists, a few enlightened policymakers, and a hardy band of environmentalists arrayed against ExxonMobil and other mighty corporations. In his account of the '90s climate deliberations, *The Carbon War*, Jeremy Leggett, a geologist turned Greenpeace climate expert, describes how he and other environmental scientists like Dan Lashof of the Natural Resources Defense Council (NRDC) first became aware of the carbon lobby. In 1988, given growing international concerns about climate change, the United Nations General Assembly set up a panel to advise governments on the issue. The Intergovernmental Panel on Climate Change, or IPCC, consisting of a large group of climate and government policy experts, was born. By 1990, after eighteen months of drafting reports on the science, impacts, and policy responses to climate change, the IPCC met in Berkshire, England, to create the final draft of its historic first Scientific Assessment Report. As Leggett tells it, he and Lashof were observers at the back of the room. Eleven scientists from the oil, coal, and chemical industries were also in attendance. The IPCC was working on a summary that noted that 60–80 percent cuts in global carbon emissions would be needed in order to stabilize the concentration of CO_2 and prevent the worst climate effects. Brian Flannery, on the payroll of Exxon and representing the International Petroleum Industries Environmental Conservation Association, intervened repeatedly. He sought to soften the conclusion that scientists were certain that global warming was related to human-made carbon emissions. He was unsuccessful in this first intervention. But more would follow, along with greater success.

Meanwhile Leggett and Lashof hoped to persuade the representatives to include clear warnings that a number of plausible projections indicated that feedback mechanisms made it possible that even greater and more rapid warming could occur. There could be nasty surprises with precedent in the geological record such as the release of methane crystals that would lead to far worse scenarios than those being discussed. They too were ignored, even though their science was quite sound. But Sir John Houghton, head of the process, who personally briefed British Prime Minister Margaret Thatcher on the problem, asserted that worst-case conclusions would be sensationalized by the press and be counterproductive. Only a vague, passing reference to possible surprises was included.[8]

By August in Sundsvaal, Sweden, Leggett, Lashof, and some other environmental representatives were at the critical IPCC session that would combine the Scientific Assessment Report that had clearly warned that climate change was happening (though with carefully balanced and politically persuasive mid-range estimates) with the reports on impacts and policy recommendations. An overall Assessment Report would be produced that would form the basis for the deliberations of the World Climate Conference in November 1990 in Geneva. It was in Sundsvaal that Leggett first saw the assembled forces of the oil lobby, including the Global Climate Coalition and the Global Climate Council. The Climate Coalition comprised, among others, the American Petroleum Institute, Amoco, Arco, Phillips, Texaco, DuPont, Shell, and British Petroleum along with the Association of International Automobile Manufacturers and the Motor Vehicle Manufacturers. Coal was represented by the American Electric Power Service Corporation, the American Mining Congress, the Edison Electric Institute, and the National Coal Association.

In addition to this grouping, Leggett first observed the machinations of Don Pearlman, a lobbyist from Washington who represented oil interests, including the Kuwaitis and Saudis, and was the head of the Global Climate Council. "As I first entered the plushest hotel in town, Don Pearlman was seated in the lobby with five diplomats, all Arab, including the head of the Saudi delegation. They had their heads down with copies of the draft negotiating text for the IPCC final report open in front of them. Pearlman was pointing at the text, and talking in a forceful growl. . . . As I walked past, I saw him pointing at a particular paragraph and I heard him say quite distinctly, 'if we can cut a deal here . . .' At the time, though it seems naïve now, I was shocked."[9]

THE BUSH WHITE HOUSE

Although the carbon lobby began to fray by the end of the Clinton administration, the election of oil man George W. Bush in 2000 gave it renewed clout. Indeed, as Ross Gelbspan, who for years has tracked the influence of the carbon lobby, points out in his second climate book, *Boiling Point,* its power even influenced the highly contested 2000 election results. Much of the controversy about 2000 focused either on the vote count in Florida or on the role of Ralph Nader. Nader drew enough votes from Al Gore in both Florida and New Hampshire to give those states to Bush. But Gore also narrowly lost in West Virginia, a poor, Democratic or "blue" state that in presidential-election years had never gone Republican. West Virginia is a stronghold of coal mining interests and influence. As Gelbspan puts it: "The Bush victory was a result of the substantial political and financial support of the state's coal industry. One coal executive alone, James Harless, raised $275,000 for Bush in West Virginia, five times the amount Gore raised in the entire state. Nationwide, coal interests tripled their contributions from the 1996 election—88 percent of which went to the Republican Party. After the Bush Inauguration, an official of the West Virginia Coal Association crowed 'You did everything you could to elect a Republican President. Now you are already seeing in his actions the payback . . . for what we did.'"[10]

The payback was soon apparent in an insider memo to the Republican Party. It was drafted by political and PR consultant Frank Luntz. He had helped create Newt Gingrich's antienvironmental Contract with America for the 1994 elections. The Luntz memo, as is often the case, only became known after it was obtained by an environmental organization called the Environmental Working Group (EWG). The memo warned that clarity on the scientific issues supporting climate change would seriously shift public attitudes about global warming. Luntz stressed: "You need to continue to make the lack of scientific certainty a primary issue."[11]

With the heavy influence of the carbon lobby ensconced inside the Bush administration, a series of shifts away from action on carbon emissions soon followed. The first was the dramatic reversal of President Bush's campaign pledge to cap CO_2 emissions. The pledge had just been reiterated by the newly appointed EPA administrator, Christine Todd Whitman, when she was quickly made to renounce it. Then the president appointed Vice President Dick Cheney to create an Energy Task Force to draft an entirely new Bush energy plan. The Cheney task force met in secret, with the direct collaboration of the energy industry and the participation of $250,000 campaign donors like Irl Englehard, chairman of the Peabody Group. Peabody, the nation's biggest coal company, mines coal not only in the United States but in countries as far off as Australia. Peabody representatives were soon numerous in the Energy Task Force and at briefings arranged by the Edison Electric Group, the lobby group for coal and nuclear utilities.[12]

But the power of the carbon lobby may be best exemplified, according to Gelbspan and others, by the influence of the merged oil giant ExxonMobil, which called for the removal of Robert Watson, an independent climate expert and chairman of the IPCC. The ExxonMobil secret memo to the White House, obtained by the NRDC through the Freedom of Information Act, pointed out that Watson had been critical of the U.S. failure to meet climate targets; he had unfavorably compared American efforts to those of the Chinese. Watson was indeed soon sacked. ExxonMobil then recommended Harlan Watson, a Republican staffer on the House Science Committee, to be head of the U.S. climate negotiating team. Harlan Watson (no relation to Robert Watson) was then appointed; he promptly announced in May 2002 that the United States would not engage in the Kyoto process for at least ten years.[13]

ExxonMobil also planned for the production of hydrogen as an energy source, relying on coal and nuclear power to create it, with no net savings of carbon emissions. And perhaps most important for public understanding of climate change, ExxonMobil became the leading source of funds for climate skeptics during the Bush years while running PR campaigns to undermine the seriousness of the global warming. By 2003, ExxonMobil gave more than $1 million a year to a number of right-wing ideological organizations opposing action on climate change, such as the Competitive Enterprise Institute and the George C. Marshall Institute.

Such efforts paid off handsomely. Oil-funded skeptics were able to directly influence conservative anti–climate change legislators like Senator James Inhofe (R-OK), who, until the Democratic takeover of Congress in 2007, was the powerful chair of the Senate Environment and Public Works Committee. To this day, Inhofe has continued to call climate change "a hoax." His industry bias was already clear in the spring of 2003, when he trumpeted a study published in *Climate Research,* a little-known German publication. Two of the coauthors, the father and son duo Craig and Sherwood Idso, were from the innocently named Center for the Study of Carbon Dioxide. But the organization's funding had shifted over time from the coal industry to ExxonMobil. Along with coauthors Sallie Baliunas and Willie Soon of the Harvard-Smithsonian Center for Astrophysics, they claimed that the twentieth century was neither the warmest nor the one with the most extreme weather events in the past one thousand years. Observed rapid warming, they claimed, was the result of solar variations. The paper was denounced by leading climate researchers (solar variations as a cause of recent warming had been already disproved by a number of peer-reviewed studies) and described as "flawed" by three editors at *Climate Research*. They resigned because they were forbidden to publish their critique in the magazine.[14]

For Americans who hope to understand the impacts of climate change in the United States, especially in their own section of the country, most telling was the carbon lobby and conservatives' assault on the "US National Assessment of the Potential Consequences of Climate Variability and Change." An analysis of the potential impact of climate change at home in the United States, the assessment is the sort of authoritative and peer-reviewed science document that, even if dense and dull, undergirds the educational efforts and reports mounted by responsible environmental groups. As Gelbspan reveals, the report was attacked by the Center for Regulatory Effectiveness, whose head, James Tozzi, "was a long-time lobbyist for Phillip Morris, General Motors and groups representing the chemical and forestry industries, among others." The critique itself was written by Pat Michaels, a climatologist and skeptic from Virginia whom I have publicly debated. Michaels has received funds from the coal industry "since the early 1990s." When the offending peer-reviewed report was not withdrawn, the White House was sued by the conservative Competitive Enterprise Institute to do so. As it turned out, in August 2003, the attorneys general of Maine and Connecticut unearthed a secret White House e-mail message to the institute requesting that it sue the administration for the withdrawal of the "US National Assessment." The instigation for the suit had come from Phillip Cooney, chief of staff of the White House Council on Environmental Quality, who had previously headed the climate program of the American Petroleum Institute.[15]

As the Democrats took over Congress in 2007, we began to learn even more about the machinations of Cooney and others close to the energy industry. Long-time environmental champion Rep. Henry Waxman (D-CA), chair of the House Government Reform and Oversight Committee, immediately summoned Cooney and others to bring forth documents and attend oversight hearings.[16] By early 2007, even ExxonMobil was beginning to change its tune, but only after years of environmentalist and consumer campaigns aimed at them. But as ExxonMobil switched CEOs and shifted its rhetoric a bit, the oil giant continued to press for domestic drilling and to side-step global warming.

The Carbon Lobby and the Fight for Clean Air

When environmentalists supported the Clinton administration on health and pollution-related issues that expose the dangers of fossil fuels, the carbon lobby counterattacked, often viciously. In 1996, when Carol Browner announced new rules to protect human health from small pollution particles under the National Ambient Air Quality Standards (NAAQS), the carbon lobby formed the Air Quality Standards Coalition. It was created and funded by groups like the National Mining Association and American Electric Power.[17] The proposed standards would save an estimated fifteen thousand preventable deaths and between $51 billion and $112 billion in annual medical costs. The cost to industry? Only about $6.5 billion to $8.5 billion. Yet, the new Republican Congress threatened to squash the proposed NAAQS under a newly minted law called the Small Business Regulatory Enforcement Act (SBREFA). It would have tied up the health-based standards in endless bureaucratic and judicial review. As sympathy for the new rules grew, the carbon lobby shifted to attacking the main academic scientist responsible for the epidemiological studies supporting the new standards. Dr. Joel Schwartz, who had begun the "Six Cities" study of air pollution at the EPA and had left for the Harvard School of Public Health, was directly smeared. The Center for Public Integrity published notes from a member of the Air Quality Standards Coalition, whose head, C. Boyden Gray, was former counsel to President George H. W. Bush: "Spent a lot of time on Schwartz. Gray spoke of impeaching his reputation, later changed to discrediting him. This crowd feels Schwartz is 'the enemy' and can be intellectually discredited."[18] A fierce battle from environmentalists and public health groups ultimately secured the NAAQS, saving some fifteen thousand lives per year.[19] Yet the net result was still further delay in capping or cutting carbon emissions and any serious prevention measures to combat global climate change.

A similar pattern can be seen in the carbon lobby's efforts to fight new regulations on mercury emissions from coal-fired electric power plants. Discussion of new rules began during the Clinton administration to restrict the output of mercury, found in coal and released into the air when burned. Mercury is then deposited into waterways where bacteria turn it into methyl mercury. It next enters plant life, is eaten by fish, and ultimately by humans. It is especially dangerous for women of child-bearing age who are or may become pregnant. Methyl mercury can pass through the placenta and the blood-brain barrier, harming the developing fetus. Exposure in utero can lead to reduced cognitive abilities and behavioral disorders among children.[20] As Jeff Goodell tells it, the coal lobby initially laid low on the new mercury ruling being proposed by the Clinton administration to take effect in 2003. Quin Shea, chief lobbyist for the Edison Electric Institute, even tried to convince the utilities represented by Edison to take a more environmentally friendly approach. It might serve them better in the long run. As Shea put it: "we stand to lose our credibility as an industry and to open the door to further attacks against coal-based generation."[21] But with the advent of the Bush administration in 2001, all this changed. The new administration folded concern over mercury into a broader "Clear Skies" proposal that did not limit CO_2 and would allow "trading" of mercury emissions so that polluters could invest in clean projects elsewhere and get credit for reducing overall emissions. Because mercury emissions, however, unlike other air pollution, fall to earth fairly close to their source, such trading would lead to mercury pollution "hot spots," many of them adding to the already heavy burden of pollution borne by low-income and minority communities.[22] Once again, environmentalists were forced to wage fierce campaigns to beat back the Clear Skies initiative.

The carbon lobby, and one of its largest components, coal, continues to have significant influence throughout the United States even as concerns about global climate change have mounted, skeptics have been exposed and proven wrong, and new, renewable technologies have become increasingly competitive. As the state of Texas was passing renewable portfolio standards to increase wind-generated power and other clean energy sources and was planning for the first large offshore wind farm, coal was moving even faster. Texas may have passed California as the nation's leading generator of carbon-free wind power in 2006, but it remains first in the production of tons of CO_2. It churns out 723.2 tons per year compared to

California's 422.3 tons and Pennsylvania's 288 tons.[23] Business and industry lobbyists in Texas have been opposing restrictions on CO_2 for some time, including opposition to a 1999 Texas House bill that would have simply required a study of Texas greenhouse gas emissions and possible steps to curb them. The same was true in 2001. Texas business lobbyists dependent on coal-powered electricity told Governor Perry that the state should keep its hand off global warming as environmentalists were pressing to include CO_2 in the annual reporting of Texas industrial emissions. Governor Perry then told the state environmental agency to do nothing until the Bush administration acted on global warming.

Coal Is Dirty

Given strong coal interests in Texas, at about the time California was passing legislation to drastically cut carbon emissions, Texas Governor Rick Perry was pushing for sixteen new coal-fired power plants. They would have added an estimated 117 million tons of CO_2 to the atmosphere each year, more than the emissions from any of 33 other states and 177 countries. A seventeenth proposed unit near Port Lavaca in Calhoun County would have burned petroleum coke, a refinery by-product similar to coal in CO_2 emissions. When TXU Energy finally announced plans to actually build eleven of the new coal-burning plants, accounting for 78 million tons of new emissions, Jim Marston, Texas director of environmental defense, told the *Dallas Morning News* that this increase was "the same amount that would result if TXU gave each of its 2.4 million retail customers four Cadillac Escalades."[24]

In order to keep environmentalists at bay, Texas coal interests (and nuclear ones as well), then said they favored futuristic solutions to reduce coal emissions and nuclear power as an alternative. In 2006, TXU announced it would spend up to $2 billion researching CO_2 solutions and would build six new nuclear reactors by 2020. Such strategies—emphasizing jobs from new coal plants, focusing on potential harm to the economy, and showing some mild concern for futuristic nuclear power and coal technologies—brought the support of some local public and private organizations, especially in the small towns where TXU was likely to build new coal-fired plants. These included, according to TXU, cities, counties, school boards, economic development agencies, and chambers of commerce.[25] Only in 2007, when a larger utility bought out TXU and recognized the intense and growing opposition to coal-fired utilities throughout Texas was the plan finally scrubbed. Local officials such as Dallas Mayor Laura Miller and Houston Mayor Bill White had joined with fifteen other Texas cities with a population of about 6.2 million people, one-third of all Texans, to fight the new coal plants with state regulators or in court. And, of course, they were backed and originally stimulated by a gaggle of environmental, public health, and consumer groups.[26] Even so, given the power of coal interests in a complex, compromise agreement, three of the planned coal-fired plants still moved forward. The rest of the power needed will come from nuclear power plants.[27]

Coal, of course, has to be mined before it can be burned. The process is dirty and devastating. In West Virginia, entire mountaintops are blown off with explosives to get at the dwindling stores of premium anthracite coal. Massey Energy is one of the leading coal mining companies in West Virginia. Regardless of the climate debate, its future depends on coal. Thus its efforts to lobby the Army Corps of Engineers to speed up the permitting process of its mountaintop-removal mines. Jeff Goodell, who has chronicled the dangers and power of the coal industry in *Big Coal,* describes how Massey advertising campaigns are designed to convince West Virginians—the public and legislators—that their economy depends on coal, even as mining jobs flee to the West where coal is more accessible in huge, mechanized strip mines. One Massey TV ad shows parents in conference with their son Johnny's teacher. The teacher is worried that Johnny is not doing well. Could there be some problem at home? You bet. Johnny's dad is being laid off because the mine can't get any more permits. They've been held up for four years. The ad closes warning of the loss of 38,000 West Virginia students like Johnny since 1990. "Support our kids, their schools, and their parents by supporting coal. Coal means West Virginia jobs at home for West Virginians."[28] Such messages are repeated wherever coal is found. Not just in West Virginia, where traditional coal mining is dwindling, but throughout the coal seams of Pennsylvania, Kentucky, midwestern Illinois and Indiana, and on to Wyoming and its huge open coal pits. Most Americans associate Illinois, for example, with corn, not coal. But even Barack Obama, despite a generally positive environmental record as a senator and expressed concerns about climate change, introduced a bill with Senator Jim Bunning (R-KY), the Coal-to-Liquid Fuel Promotion Act of 2007, to subsidize research and development of coal liquefaction technologies that would help the coal industry in southern Illinois and Kentucky. But it would not reduce climate change emissions.[29]

As with Big Oil, Big Coal's newfound seat at the White House policy table led it to oppose any regulation, whether of adverse environmental effects, of safety issues related to coal mining, or of CO_2 and other pollutants pumped into the air. With the election of George W. Bush, the Clinton-appointed head of mine safety at the Mine Safety and Health Administration (MSHA), J. Davitt McAteer, a safety expert, was replaced by Dave Lauriski. Lauriski, a former coal executive, pressed to have the mining industry police itself. His boss, Secretary of Labor Elaine Chao, was the wife of Senator Mitch McConnell (R-KY). As head of the National Republican Senatorial Committee, he had received $584,000 for election work from the coal industry. The influence was soon felt. According to Jeff Connell, in 2002, Ohio mine owner Bob Murray threatened to have some MSHA inspectors fired who had cited his mines for violations. "Mitch McConnell calls me one of the five finest men in America," Murray boasted to the inspectors, "And last time I checked, he was sleeping with your boss."[30]

The influence of coal goes far beyond such crude bullying over mine inspections, however. Because coal is hauled by freight trains, an iconic American symbol, the railroads, too, have joined with coal companies to fight restrictions on coal mining and on carbon emissions and climate change efforts. Matt Rose, the CEO of the Burlington Northern and Santa Fe Railway (BNSF), for example, was, in both 2000 and 2004, a member of the Bush Pioneers, a group of fundraisers who must give or bring in $100,000 to qualify. BSNF was a member of the Global Climate Coalition that fought the Kyoto treaty and the Greening Earth Society that promotes the idea that global warming is good. Marc Racicot, a BSNF board member has also been chairman of President Bush's reelection campaign and chairman of the Republican National Committee. John Snow (not to be confused with the activist London physician), secretary of the treasury, is the former head of CSX Corporation, a Florida-based railroad that also hauls coal.[31]

Undue Influence

If the influence of the carbon lobby has been substantial in shaping the scientific approach, public perceptions, and policymakers' attitudes toward climate change, the centrality of oil, natural gas, and other energy resources to U.S. foreign and military policy has been a similar obstacle. The Bush administration has, of course, been the most extreme example of the influence of oil and other energy interests in government. Not only was the president himself an oilman (like his father George H. W. Bush), but Vice President Cheney, as is now well known, was CEO of Halliburton, a huge oil services corporation. Secretary of Commerce Donald L. Evans worked for Tom Brown, Inc., an oil and gas company, and Secretary of State (and former National Security Advisor) Condoleezza Rice served on the board of directors of Chevron. Andrew Card, the first White House chief of staff, is a former president of the Automobile Manufacturers Association. Former Interior Secretary Gale Norton had received more than $285,000 from energy industries for her state senate race in Colorado and then chaired the BP Amoco– and Ford–funded Coalition of Republican Environmental Advocates, while Secretary of Energy Spencer Abraham was appointed after losing a U.S. Senate race in Michigan where he had received more than $700,000 from the auto industry.[32]

Given the connections between climate change and some of the most powerful interests in the United States, if Americans and a new, broader environmental and climate movement are to be successful, we will need to find more successful ways to think and talk about—or frame—global climate change and its serious impacts already becoming visible around the world.

4

Framing and Talking about Global Warming

Global warming is not only a growing and dangerous public health problem. It is deeply embedded in the nature of the American economy, with strong corporate and political lobbies that seek to minimize its importance and forestall change. Simply presenting the "facts" will not be sufficient to counteract the carbon lobby. Even with a new administration in 2009, to have hope for a real transformation of American society—its energy use and greenhouse gas emissions—sustained political engagement from the public in favor of new, positive initiatives will be needed. To achieve this, Americans far beyond the usual circles of scientists, environmentalists, and policy wonks need to be involved.

As with other public health issues, how the dangers and solutions are defined and discussed shapes how the problem is perceived. With global warming and environmental issues, recent criticisms of the mainstream environmental movement, such as "The Death of Environmentalism," by Michael Shellenberger and Ted Nordhaus, have accused traditional, large environmental groups of being too cautious and narrow. The movement to stop global warming is not bigger, or stronger, it is asserted, because big green groups are guilty of mere incrementalism. They mount campaigns built around narrow legislative goals like CAFE (Corporate Average Fuel Economy) standards. These efforts have neither been successful, nor have they given rise to the visionary leadership, moral passion, and popular enthusiasm needed to combat climate change. Even focusing on an entity called "the environment," it is said, is off the mark.[1]

While offering fresh thinking about the need for new frameworks, political alliances, and movements, such critiques seem to imply that an entirely new movement, freed from the constraints of old environmental organizations, is the only way to proceed. They come perilously close to throwing the baby out with the bathwater. They have not sufficiently appreciated the shift to grassroots organizing, the alliances with new constituencies, and the reframing of climate work in terms of human health, community well-being, moral values, national security, and positive, far-reaching energy solutions that have already changed green organizations. Groups such as the Natural Resources Defense Council, the Union of Concerned Scientists, Environmental Defense, the National Wildlife Federation, and others had all begun to shift tactics and expand their allies long before they received already outdated public rebukes.

A report, for example, from Clear the Air, a coalition formed by mainstream environmental groups to battle air pollution, relates to climate change, the group's current top priority. But it features up-to-date medical and public health studies. These link carbon emissions and pollutants from coal-fired power plants to new evidence connecting air pollution with lung cancer and heart attacks. The study is, of course, about the environment and climate. But it is essentially a moral and medical appeal; it is about people like you and me.[2] The roots of this change in approach go back over a decade when the Green Group had already decided to expand its horizons and include, among other new groups, Physicians for Social Responsibility (PSR).

CLIMATE: THE HEALTH FRAME

For Physicians for Social Responsibility (PSR), which first set out in the early 1990s to organize the medical community and inform the public about the health impacts of environmental toxins and climate change, some key insights were gained about the relationship between the environment and health. These often came from advances in modern medicine, genetics and public health, cognitive science, and communications.[3] Now, thanks to George Lakoff and others, activists and scholars alike call the process of choosing how we present, describe, and communicate our ideas *framing*. As Lakoff has famously put it, if you ask people not to think about an elephant, they will.[4] The human brain—and the imagination, emotions, and language intimately connected to it—is not a tabula rasa. Our minds utilize preexisting frames, wiring, metaphors, pictures, stereotypes if you will, to speed insights and sort out the irrelevant. In political and policy discourse, these preexisting notions can be tapped and used, manipulated through advertising and repetition. Environmentalists? Tree huggers. Elitist backpackers through wilderness and parks more concerned with vistas and vacations than with you or your job. More about saving the snail darter or the Snake River than saving the local sawmill. More about Kyoto than coal miners. These are, of course, false stereotypes, even vicious ones. But to some degree they flow inexorably, like runoff to the Chesapeake, from objectifying the environment as things in nature, landscape, species other than humans. They exist outside of and apart from us. The environment is to be observed, enjoyed, measured, and protected. It is hard, though not impossible, given such a standpoint, to put people and core human values back into the picture. As Lakoff puts it in *Moral Politics* in discussing how to reframe the Bush administration's emphasis on an energy crisis and the need for more supplies of energy, no matter the consequences:

> Energy is part of a system, an ecological system. Energy is not an isolated issue. It is a *Health* issue: the purity of the air we breathe, the water we drink, and the food we eat are all Health issues. How air smells and water tastes are *Quality of Life* issues, as is the role of nature and beauty in our lives. Energy is also a *Moral* issue. How we get our energy and whether our government and businesses allow for its efficient use so that we minimally despoil the earth are questions of morality. We have a responsibility to take care of the Earth, a religious responsibility if you believe in God and an ethical one if you do not. [Italics in the original.][5]

That is why PSR, working with other environmental and public health groups—Environmental Defense, the Environmental Working Group, the National Environmental Trust, and the Natural Resources Defense Council, as well as grassroots coalitions and organizations like Citizens for Environmental Health and Justice, Health Care without Harm, and community groups from Harlem to the US-Mexico border—began to develop a different frame for the environment. Working collaboratively, PSR instituted campaigns and programs in which human health was the primary frame for the control of toxic chemicals, the promotion of clean air and water, and the support of international treaties like the Stockholm Convention to ban the twelve most dangerous persistent organic pollutants (POPs) and, of course, agreements for the prevention of global climate change.

The environment is *not* simply something outside of us. Cactus and vermillion flycatchers. Coyotes. Caribou. Don't get me wrong. The Arctic

National Wildlife Refuge is a very real place, and the Bush administration wants to spoil it by setting up real rigs there to get oil. The permafrost in the Arctic is also quite real—even if it is now slipping out from beneath buildings thanks to global warming that has increased average temperatures at the top of the globe by some 4–8°F.[6] But the environment is also *inside* us. It *is* us. Human health studies of the Inuit living in the Arctic circumpolar region reveal that their bodies contain accumulations of PCBs, mercury, dioxin—minute quantities of toxins whose molecules have traveled thousands of miles to get inside their bodies.

Now we know that these molecules, carried by wind and deposited in water, enter the ecosystem and move up the food chain into fish and seals and whales and are then eaten by the Inuit. The result can be disruption of the endocrine system, learning and developmental disorders, low birth weights, even cancer.[7] These are relatively new insights about what happens to us as what we call the environment moves freely into, throughout, and back out of our bodies. Such new understandings depend on rapid advancements in microbiology and biomedical engineering along with countless epidemiological studies of populations of animals all over the map—alligators, herring gulls, beluga whales, human beings. Our ability to see and understand how the dumping of an electrical lubricant into a landfill in Texas can be found in and affect the cellular function, the health, of a child in the Faroe Islands is a revolution in our thinking as profound as the photographs from space that first allowed us to see ourselves as part of one fragile, finite planet—a global ecosystem.[8]

Our ability to observe and measure the effects of, say, mercury, a potent neurotoxin from the environment inside our bodies, our cells, our brains, has given rise to an environmental health movement that continues to grow in size and influence.[9] The blurring of distinctions between the outside environment and the inner environment of the human body has allowed the frame for this new movement to shift from primarily caribou or cormorants (whose bills, in the case of endocrine disruption, become malformed and twisted) to the health and well-being of people. It allows activists to offer concern and ultimately hope for healthy homes and communities, families with disabilities, workers in farms and factories, women who may become pregnant, urban joggers and cyclists, and more. The environmental health movement has already achieved important victories such as a ban in the European Union (EU) of obscure plastic softeners called phthalates in baby toys. The scientific, philosophical, and communications boundaries between a baby toy as a solid, safe, practical item *out* there and what enters a baby's body at a molecular level *in* there has been breached.[10] The same is true of the removal of the phthalate plasticizer DEHP and polyvinyl chloride plastics (PVCs) from lifesaving and life-giving drip bags and other plastics in hospitals, the incineration of which has been a major source of the dangerous organic compound dioxin. It is the same dioxin that can travel all the way to an Inuit village.[11]

Reframing environmental questions in terms of human health has the benefit of involving everyone across boundaries of neighborhood, class, ethnic group, or nation. Such a shift in focus also reveals that our environmental choices are often shaped by and exacerbate long-standing divisions in the United States based on class and race. Protection of the environment in this country was originally a concern mainly of the upper class that sought to preserve wilderness, scenic beauty, recreation spots, and resources that they considered essential to the good life. But in recent years, pushed by a growing environmental justice or environmental racism movement, mainstream environmentalists have begun to work to spotlight the effects of pollution and climate change on the most vulnerable Americans. These include racial minorities who live closer to many sources of industrial, automobile, and chemical pollution, and who often lack adequate medical services and care; children and the elderly; and those, who, for a variety of reasons, have compromised immune systems.[12]

Shifting the frame to individual, community, and group health concerns also adds a powerful set of respected messengers—health professionals—to environmental debates. Physicians, nurses, and public health scientists remain credible spokespeople in an age where almost every group—especially the media, the White House, Congress, and other politicians—has seen a decline in trust and credibility.[13]

The Health Link from Climate to Air Pollution and Energy

Even though organized groups of public health professionals interested in environmental issues are not large and do not seem to offer huge numbers of participants as do labor unions or churches, they have already proved themselves to have disproportionate influence. This was particularly clear in the great anti-nuclear campaigns of the 1960s and 1980s, when physicians added professional, scientific, moral, and community standing to a movement that faced even more severe stereotypes and ridicule than the modern environmental movement has ever seen. I have personally experienced many times how even one well-regarded physician can get me a meeting and a serious hearing with almost any senator or congressional representative. Active, concerned physicians often have been engaged in community or family medicine, public health, and in grassroots coalitions. They recognize, more than many professionals, the power of the health framework. The best health professionals are used to expressing the concerns of Americans for the health of their families and communities and in engaging wider constituencies. The links between clean air and health, and attempts to prevent toxic chemical exposures have already built an environmental health movement in the United States with real grassroots strength and sophisticated links to national groups, coalitions, and policy change. Although it is not simple, drawing the connections between long-standing concerns over air pollution and toxic chemicals offers a particularly fruitful way to broaden the environmental constituency, messengers, and message.

Air Pollution and Health

In the United States, the vast majority of our energy comes from the combustion of fossil fuels—oil, gas, and coal. This simultaneously produces CO_2, the leading greenhouse gas, and a variety of pollutants known to harm human health. This link between energy production, carbon emissions, air pollution, and human health offers the chance to gain immediate benefits on two fronts. The medical and scientific connections are well established; they have been a concern for society since coal was first mined and widely used as a fuel in the time of Edward I's thirteenth-century England. It was Edward who promulgated the world's first air pollution and public health regulations, banning the use of coal in London. By the time of the invention of the steam engine and the dramatic leap in coal use in the Industrial Age beginning in the 1750s in England, the unhealthy effects of coal burning were understood; the danger of releasing tons of CO_2 into the atmosphere was not. Thus, until the late twentieth century, concern over pollution from fossil fuels (oil and gas were added to the mix in the late nineteenth and early twentieth centuries) and human health was unconnected to global climate change.[14]

Public concern among public health scientists, physicians, urban planners, and (later) environmentalists grew slowly and finally reached national and international proportions after the killer London fog of 1948. It left five thousand dead in a few days, as did similar killer fogs in Donora, Pennsylvania, and elsewhere.[15] These catastrophes and a growing environmental movement led to the Clean Air Act of 1970 in the United States. It mandated a clear health standard based on the best medical and public health evidence, reviewed every five years. Thus, when the climate effects of the "fourth" air pollutant, CO_2, became better known in the last quarter of the twentieth century, the environmental movement's concern with air pollution and a regulatory framework based on human health, not cost or technological feasibility, offered a powerful tool for change. By the 1990s, a small but growing number of physicians and other health professionals began to link uneasiness over climate change to established environmental concerns like air pollution.

Death and disease from dirty air, which is directly related to and exacerbated by global warming, currently accounts for more than fifty thousand deaths annually in the United States. The bulk comes from vehicle exhaust and coal-fired power plants.[16] The main culprits are increased particulate air pollution (soot) from sulfur dioxide and ozone (smog) from volatile organic compounds (VOCs). Environmental groups were well aware of global climate change and warned about it during the 1990s. But they often found it difficult to organize around. The only legislative votes available were on treaties. Very local effects in the United States were hard to document. Because movements are not built in the abstract, the environmental community took on battles at least related to climate change that would engage ordinary Americans.

PSR was central to a huge struggle over new, tougher National Ambient Air Quality Standards (NAAQS) proposed by Carol Browner's EPA. Environmental groups had been suing over the long-standing failure of the EPA to update the regulations. They are supposed to be based on regularly reviewed health studies and analysis as required by the Clean Air Act. It is the only environmental law that requires adherence to health standards regardless of the cost or convenience to industry.[17] It is, in fact, why the new rules were fought so hard by energy producers, auto manufacturers, auto unions, and their congressional allies. Faced with yet another huge, conservative backlash against regulations that would save fifteen thousand lives a year, PSR joined with the National Environmental Trust (NET), the American Lung Association, and others in a national campaign to save the NAAQS from being overturned by the Congress.

In a classic grassroots campaign featuring a 15-foot high, theatrical-size tombstone resembling Ebenezer Scrooge's worst nightmare, NET and PSR traveled to key congressional districts, joined with local coalitions, asthma sufferers, pulmonologists, and moms, gaining massive publicity and impact on legislators *at the grass roots.*[18] A defining moment came when a new medical study indicated a correlation between Sudden Infant Death Syndrome (SIDS) and particulate air pollution. A hastily called national press conference with environmentalists, physicians, and organized mothers who had lost babies to the mysterious and dreaded SIDS was perhaps the final blow.[19] Congress backed off trying to repeal the new NAAQS. Some fifteen thousand American lives were saved. And, importantly, alliances at the local level, with the medical and public health community, and with environmentalists were forged and strengthened. At PSR, the very same members, activists, and allies gained through such efforts also respond to calls to combat global climate change. They see them as inextricably linked and opportunities for success, not mere narrow and incremental Band-Aids. Integrated, yet pragmatic, campaigns around air pollution, human health, and climate change do bring a unifying vision and moral passion to environmental organizing. Homes and human life are at stake; the moral concern of physicians to prevent harm and to take precautionary, preventive action becomes central.

The Mercury Campaign and Framing the Climate Message

A prime example is the environmental campaign against the Bush administration's mercury rule. On the surface it seems small, arcane, legislative, incremental—one more losing Beltway battle. And it doesn't address climate change. Well. Wrong. The mercury rule proposed in December 2003 by the Bush administration's EPA to control toxic mercury emissions from coal-fired utilities was indeed obscure. It reduced mercury emissions overall but did not target specific plants, and it allowed polluters to purchase credits for cuts made elsewhere. Worst of all, the rule put off until 2018 cuts from 48 tons of mercury being emitted in 2003 to 15 tons—still a very dangerous level. Why mobilize against a single pollutant and a complicated rule when global climate change threatens the whole planet? The mercury campaign must be seen in the larger context of the administration's assault on decades of environmental gains and its failure to act on any issues, including global climate change, essential to the environment and human health. Besides, movements need victories to sustain themselves. The mercury rule was under immediate consideration. It was extremely harmful to real people. It called attention to the connections between local power plants, pollution, health, and climate change. And, it might be stopped.

It was, in fact, years of renewed grassroots organizing, media work, and campaigns by environmental and health groups that had begun to reframe their approach that gave rise to a massive outcry and outrage over such an obscure rule. After PSR revealed that the EPA had ignored its own expert panel on mercury, the Children's Health Protection Advisory Committee, and leaked a committee letter to then EPA Administrator Mike Leavitt stating that the proposed rule "does not adequately protect children," all hell broke loose. PSR further made the inadequacy of the rule clear by reviewing and touting carefully buried EPA pronouncements. They revealed that more than 630,000 newborn infants per year (far higher than generally supposed) were exposed in the womb to unsafe levels of mercury. Environmentalists, public health groups, mothers, disability activists, and consumers were now fully banded together. More than six hundred thousand individual public comments were drafted and sent to the EPA on its proposed rule. More than twelve thousand of them were from the medical community, as well as from forty-five U.S. senators, ten state attorneys general, and numerous state and local officials. PSR then led national press events in March 2004 with a coalition of medical, nursing, public health, labor, learning disability, and faith groups. All spoke against the mercury rule. By linking it to both clinical and consumer guides to fish consumption and how women of childbearing age could limit exposure, PSR was also able to penetrate popular food and women's magazines, health Web sites, and journals. With a video press release featuring Erica Frank, MD, of the Emory School of Medicine's Department of Preventive Medicine, the campaign got coverage on TV in 85 media markets with 162 stories reaching 10.5 million viewers.[20] Such campaigns, rather than showing the failure of nerve or a lack of moral passion and vision on the part of Beltway organizations, indicate the power of putting traditional environmental issues and climate change within a community and human health framework.

There is no silver bullet for preventing global climate change, but putting people and their health, safety, and well-being first helps. The Americans we must talk to are not blank computers waiting to receive data, nor is it effective to urge them to worry more about whales if they are worried about their families. Often, they will "frame" our work for us, if we will but listen. On one speaking trip to Maine, I found myself sitting in the office of a major developer, seeking funds for PSR's work. I had been brought there by Peter Wilk, MD, PSR's past-president and head of our PSR/Maine chapter. The conversation rambled through PSR programs and priorities until I asked about our potential donor's children. I was taken aback at the sudden pride and passion as Peter's friend talked about his asthmatic son and how deeply he cared about him; how outrageous it was that their home was near an EPA monitoring station that steadily revealed hazardous levels of ozone. He knew, thanks to the work of health professionals and environmentalists, that ozone is a cause of and exacerbates asthma attacks and that things get even worse with increasing heat and, ultimately, global warming. "If you can do something, anything, serious about that, I will give you more money than you are talking about now, and I will engage others," he said.

Energy Solutions

The same holds true for the environmental movement's framing of global climate change as an issue that requires positive, innovative new energy solutions that will create a healthier and a more technologically advanced and competitive U.S. economy. As we organized across the United States, both medical professionals and citizens began to say that they had been convinced that climate change was indeed real. They agreed that such serious disruption of climate systems would be harmful to human health, their families, livelihoods, and the nation. What they wanted to know was what PSR and other groups had to offer as solutions—in policy and in their personal lives. This frequent request coincided with a framing exercise carried out for the Green Group organizations by the Frameworks Institute. The findings, boiled down, were that environmental organizations should explain the science of climate change in simple, human terms. Most important, they should offer solutions to such an overwhelming problem. When caricatured, such an approach can seem ludicrous in the face of planetary catastrophe. Standing alone, advice such as "Go buy an energy-efficient refrigerator and new light bulbs" neither inspires nor fully solves the problem.

I had, in fact, initially opposed such suggestions as part of PSR's approach. They seemed to put the burden of action and guilt too much on the individual, while falling far short of, and possibly undercutting, the drastic reductions required. But what was being suggested by the Frameworks Institute, and what I came to appreciate in traveling throughout the country, was the effectiveness of a positive, practical appeal. Leave behind the familiar fear, gloom, and doom. It often leads to despair and disempowerment. Bringing the lofty stuff of climate science down to the lights in your house is a way to engage individuals, to give them choices, to meet their needs and concerns, and to educate them as well. It does not mean that average Americans are then somehow deterred from grasping the big picture. It is a way for them to enter *into* the picture. As much as I know about climate science, the environment, and health, I had never bothered with "trivial" things like improving my light bulbs. That is, until my organizing staff insisted on including them in PSR climate reports. By then, I had seen at climate change and clean air gatherings enough vivid examples of the dangers of electricity production to be convinced. My parents had always admonished me to turn off the lights when leaving a room (as have members of my own family). But what did it for me, and changed the bulbs in my home to compact fluorescents, was a huge picture of a 500-pound pile of coal I saw in a classic church basement meeting. That is roughly the amount of coal that is burned and goes up into the sky over the life of a single, old-fashioned incandescent light bulb. There is no contradiction, I believe, between individual, small actions and incremental political battles and victories where you can get them, and a larger, global and moral vision. Addressing issues that are relevant in people's and politicians' lives is not selling out. Big environmental organizations have often dwelt on the energy bills in Congress (and their complicated subparts). But these actually will determine what much of our energy future will look like. In lobbying, say, for renewable portfolio standards or more subsidies for wind power, big green groups are, in fact, mobilizing more people and getting them into the climate battle, rather than keeping them from it. Energy and how it is used is central to all our lives in the United States. In talking about light bulbs and liquid natural gas or the growing use of wind and solar power, major environmental organizations have not been neglecting climate change. They have learned, through experience and research, that we Americans are indeed a pragmatic bunch, as well as principled. We want to know what will work.

Starting with the primary season leading up to the 2004 election, PSR sought to incorporate these lessons into our public health approach to climate change. PSR's second climate campaign, called "A Breath of Fresh Air," featured reports and organizing that defined the climate issue in terms of how Americans can build a new, vibrant, and sustainable economy for the future. How can we end the outmoded and unhealthy fossil fuel age that is causing global climate change and hazardous air pollution while taking direct action to protect our families, friends, and community?

In Pennsylvania, for example, PSR's approach and *Breath of Fresh Air* report was endorsed by PennFutures, a statewide coalition and lobby concerned with energy efficiency and renewable energy. It was also backed by the Pennsylvania Council of Churches and local environmental groups.[21] And, as was the case with the *Death by Degrees* reports, it has engaged a statewide network of medical and public health experts and health officials, medical schools, hospitals, and universities. Its method combines the passion and vision of the Hippocratic Oath with the stewardship concerns of the faith community. It creates an overarching vision of human health and wholeness. It offers both hope and action to protect our children and their future. Within such a frame, the health effects of Pennsylvania's aging coal-fired utilities, the production of mercury, the need for fish and health advisories, policy advice on global climate change legislation, and information on business and individual choices and investments in solar, wind, and energy efficiency cohere in an integrated, powerful approach.

In such a context, anglers, academics, the American Federation of State, County, and Municipal Employees, the American Baptist Church, and the American Heart Association can all be involved. Framing the issue for all Pennsylvanians in terms of human health and community makes sense of and links together some simple and startling statistics. In Pennsylvania, fine particle pollution from power plants alone is responsible each year for 35,405 asthma attacks, 3,329 heart attacks, 194 lung cancer deaths, and 200,100 lost workdays.[22] As John F. Kennedy said in announcing an end to open air nuclear testing on June 10, 1963, following years of widespread education and protest from the anti-nuclear movement, the world's first truly global environmental health and human rights campaign of which PSR was a key part: "We all love our children. We all breathe the same air. We are all mortal."[23] A frame of human health and positive, practical solutions had helped bring the abstract threat of global nuclear war and radiation poisoning down to ordinary citizens in the United States. The combination ultimately vastly reduced nuclear weapons and the threat of nuclear war.[24] Such can be the power of framing global climate change in similar terms of healing for humanity.

Energy, Design, and Sustainability

As the public health community and environmentalists have begun to present climate change through energy solutions and human health, other ways of presenting climate change in new contexts or from fresh angles have also emerged. One that grows out of concerns for energy use and health is the huge, but exciting, prospect of redesigning the entire American infrastructure or built environment to create new, sustainable communities. Few of us think about how the way in which our hometown or city is designed and laid out, or how its roads and buildings are constructed, is related to climate change. But already coalitions of business, environmentalists, the medical and public health community, urban planners and designers, architects, developers, bicyclists, and other recreationists are involved. They are joining networks such as SmartGrowth America or Clean Air/Cool Planet in the Northeast that, as we will see in chapter 6, are part of a rapidly growing new climate movement.[25]

I was particularly impressed with the potential of discussing climate change in terms of sustainable energy use and healthy building designs on Earth Day 2005 at Harvard's Longwood Medical Campus, an urban district that houses the Harvard Medical, Public Health, and Dental Schools alongside some of the most prestigious medical centers in the world. A day-long symposium at which I spoke, along with Eric Chivian, head of the Harvard Center for Global Environment and Health, brought together public health and medical experts, business representatives, designers, campus facilities staff, students, faculty, and citizens to discuss and act on Harvard's new sustainability initiative.[26] Similar events are occurring more and more across campuses nationwide, including less-visible campuses such as Ithaca College in upstate New York. It has hosted a New York State sustainability summit where I spoke on the health benefits of sustainability alongside business leaders, academics, environmentalists, urban designers and gardeners, faculty, and staff. Such events educate, organize, inspire, and create real solutions—all while showcasing the college's own sustainability plans. These range from a new, entirely green building for its business school, donated by a trustee, to food from its award-winning EcoVillage and organic gardens where the students and local residents work together to meet the needs of the college and the wider Ithaca community.[27]

Local and State Solutions

Obviously, reframing the environment means changing and widening the definition of who is an "environmentalist." This lesson was brought home to me vividly when, thanks to a PSR *Death by Degrees* report in Michigan and the inclusion of a respected Michigan doctor who had participated in the report and press events, I was able to meet directly with the Republican head of the Michigan Senate Environment and Public Works Committee. Our report had already provoked a statewide media response from the anti-Kyoto Chamber of Commerce (thus increasing the reach of our message many times over!) before my meeting. Initially, the senator let me know why Kyoto was unfair, why he opposed the treaty, and why there was little he could do for me. But as we talked about the health impacts of climate in Michigan and, specifically, the increased amount of precipitation as heavy rain and storm sewage overflows that would dump increasing amounts of human wastes loaded with *E. coli* bacteria into Michigan water, we became allies. We left Kyoto behind. We happily began to plot how he could steer appropriations into storm sewage control, help human health, and stave off some of the effects of climate change with one mighty stroke of his pen.

As PSR and others have moved away from a primary emphasis on environmental threats to solutions to protecting human health and the ecosystems that support us, a new movement has been building and finding increasing success at state and local levels. Our work in Pennsylvania, along with many partners, led to new Renewable Portfolio Standards (RPSs), jargon for state requirements that a percentage of electricity production come from renewable sources such as wind and solar power. The same is true for many other states where business interests, environmentalists, and medical and public health professionals have combined under the banner of jobs and healthy communities. In Maine, PSR's original *Death by Degrees* report engaged the former Commissioner of Health while giving fresh arguments and allies to environmentalists. They have since mounted state conferences, lobbying, and actions that have affected the state legislature, Maine's two important moderate Republican senators, and more. Such efforts linked with local and regional climate and environmental groups like Clean Air/ Cool Planet, the Maine Natural Resources Council, and others also laid the groundwork for the election of environmentalist Governor John Baldacci.

Energy and National Security

Another new approach that combines concerns for global climate change, human health, safety, and well-being is what is coming to be called energy security. This frame suggests our ability to enhance U.S. national security and protect American lives and families by reducing dependence on oil and gas. The nation's strategic dependence on petroleum, as we saw in chapter 3, has led to military and foreign policy strategies designed to secure access to oil and gas through U.S. involvement, intervention, deployment, and basing in places like Uzbekistan, Iraq, Saudi Arabia, and Nigeria. At the same time, the United States can reduce its vulnerability to terrorist attacks on hazardous and centralized sources of fuel and power such as oil refineries, pipelines, power stations, tankers laden with liquefied natural gas, nuclear power plants, waste storage sites, and the roads, rails, bridges, and tunnels that lead to them. Needless to say, such a shift in focus brings together unlikely allies and underscores immense human health benefits related to the pollutants produced by fossil fuels.

An early Capitol Hill briefing in the U.S. Senate sponsored by PSR and the Heinrich Böll Foundation offers a good example. Former CIA Director James Woolsey, a self-proclaimed "cheap hawk," talked about the need to move away from fossil fuels toward renewables and efficiency as quickly as possible both to protect U.S. security and to prevent global climate change. In agreement with Woolsey about the health, economic, and security benefits of ending fossil fuel use were Susan Tierney, chair of the Energy Foundation and a member of the National Panel on Energy, and Kent Bransford, MD, PSR's senior medical consultant on energy and climate. Bransford not only outlined the health benefits of new energy sources, but also drew a warm response from both sides of the aisle and the ideological divide when he suggested that it made no sense to send American troops in gas-guzzling, poorly armored vehicles to Iraq to protect access to oil so that Americans could have the right to drive gas-guzzling SUVs and Hummers here at home. Simply increasing CAFE standards by 2 miles a gallon would eliminate the need for Iraqi oil entirely while protecting human health and staving off global climate change at the same time.[28] Again, the frame of the energy/climate/health/security nexus meant that CAFE standards are not simply some incremental, poorly thought-out environmentalist dream that has backfired. Instead, Republicans and Democrats, hawks and doves, energy specialists and environmentalists find a common vision and common ground.

Moral and Religious Vision

But what about the grand vision, moral principles, even religious zeal? Damn it, God's creation is at risk; don't just give me light bulbs, local organizing, and lectures on oil dependence! Here, too, the environmental movement has been deepening its moral and religious base for some time. It has been reaching out to and connecting with faith-based groups and letting other Americans know that they share their fundamental values. To mess with God's creation in ways that threaten every living thing in it is simply *wrong*. To destroy the earth violates our most deeply held beliefs and cherished values, however we express them. Environment and ecology are, indeed, scientific subjects. But they are not only that. For more than a decade major environmental groups have been lining up alongside religious ones to say to Americans, of both secular and spiritual persuasion, that both approaches to life share something in common. The Green Group had included the National Religious Partnership for the Environment (NRPE) in its ranks long before Kyoto. And, as early as 1997, Carl Pope, executive director of the Sierra Club, had urged stronger alliances with faith-based communities and publicly apologized for some environmentalists' overly negative assumptions about the role of Christianity and Judaism in giving humanity license to plunder the earth.[29] As Paul Gorman, the wise and witty leader of the National Religious Partnership told me, he could not understand why there weren't bigger headlines reading: "Pope repents!"[30] By 2005, even the leader of the IPCC Scientific Assessment, Sir John Houghton, a renowned climate scientist who had personally briefed and convinced British Prime Minister Margaret Thatcher about the problem, spoke to an evangelical Christian gathering as both a scientist and a concerned Christian.[31]

What the environmental movement and its critics do agree on is that to date the approaches used by the mainstream core of the environmental movement have not been sufficient in the face of determined corporate and political opposition. We still need to create a broader, deeper, more inclusive, much larger, and more sophisticated movement. Additional organizations, initiatives, and some real passion will be necessary, as will new ways to present or frame environmental concerns. The good news is that much of what has been called for in recent critiques of the mainstream environmental movement had already been developing at both national organizations and coalitions and in local communities and grassroots groups for a long, long time. And, it is working. Have hope. The ark is not fully built, but it is well under way.

5

Assessing the Big Greens: Start Over or Build from Strength?

Framing environmental issues and climate change so they appeal to the broadest possible segment of Americans is critical to building support to prevent global warming. Powerful economic and political interests remain arrayed against change. Only public opinion that is mobilized through public interest organizations can offer a sufficient counterweight. But groups must be respected, strong enough, and actively engaged in our nation's political process. A detailed public health strategy evaluates the organizational and political forces available to it. In this chapter, before turning to newer, still emerging movement efforts, I will examine closely the mainstream environmental movement, especially the informal coalition called the Green Group, its most serious critics, its frail beginnings during the George H. W. Bush administration, and some of its achievements in the Clinton years. These efforts fell short of stopping climate change. But they gave us the historic Kyoto Protocol; they laid the foundation for the growing movement and renewed interest in climate change we are witnessing today.

Back in 1990, as we saw in chapter 3, Jeremy Leggett of Greenpeace and Dan Lashof of the Natural Resources Defense Council (NRDC) were lonely figures in the back of a conference room in Berkshire, England. They watched Don Pearlman and others from the carbon lobby work the room to soften the first IPCC Scientific Report's conclusions. Sir John Houghton, the chair of the group, seeking consensus and credibility, counseled a moderate approach.[1] Despite its tempered tone and findings, the first IPCC climate assessment convinced Margaret Thatcher, the conservative prime minister of Great Britain, personally briefed by Houghton, that global climate change was a threat. But in the United States, the moderate conclusions strategically adopted by the IPCC report were not the stuff of headlines. It failed to provide enough grist for the media mills or for environmentalists to effectively mobilize Americans; it was ignored by the first Bush White House. A decade-long battle was about to begin, mostly out of public view, to alert both average Americans and policymakers to the dangers they were unable to see.

Waged by a few environmental groups and climate change pioneers, these early skirmishes have been largely forgotten, or glossed over in ways that suggest that the Kyoto treaty was not a significant achievement. Then, as Al Gore and John Kerry came close to defeating antienvironmentalist George W. Bush, the environmental movement, especially its larger national organizations, took much of the blame. They must lack vision, daring, militancy, diversity, an understanding of social change, and the plain old working-class chutzpah necessary to get the job done.

Yet if the mainstream environmental movement is so flawed, if social change only comes with dramatic mobilizations, direct action, and diverse constituencies, how did an essentially middle-class movement of reformers, not radicals, manage to gain a global treaty and convince a cautious, besieged administration to sign it? How did such supposedly flawed presidential candidates as Al Gore and John Kerry, backed by environmentalists, come so close to winning? The answer is of more than academic interest.

The Big Greens under Fire

I have already talked about how reframing environmental issues can help build a bigger movement. I have also argued that calls for an entirely new climate movement, seemingly devoid of the old groups and their limitations, are off the mark. It is impossible to organize practically *de novo;* it is dangerous to airily debunk organizations that got us this far. We will need the current environmental movement, especially its organized and institutional parts, as an essential, though not sufficient, part of the push for progress.

The environmental movement has become in recent years the whipping boy of those frustrated by the lack of gains before and after Kyoto. Most have been well intentioned; their criticisms have caused the environmental movement to seek greater unity and vision, media savvy, and political acumen. But, as is the case with progressives, they sometimes point harsher fingers at those who have helped to lead reform movements (and moderate and liberal politicians who work with them) than at the real culprits.

To date, none of the growing shelf of analyses of the environmental movement has been especially kind. Some are decidedly hostile. Most save their toughest hits for the much maligned "mainstream" movement. It is caricatured as composed of compromisers trapped inside the Beltway. I have been a participant in and friendly observer of the mainstream environmental movement, a leader in the reform wing of the American peace movement, and a national producer and reporter for alternative radio designed to reach millions of Americans who might be influenced by reasonable programming about war, peace, and the environment. As head of Physicians for Social Responsibility (PSR), I worked regularly with the Green Group of thirty-four national environmental organizations that cooperate in Washington, D.C. The Green Group is an informal coalition. It tries to plot, coordinate, and carry out strategy and tactics to move environmental policies forward. In recent years, it has successfully defended most of the gains made since the 1970s and the "first wave" of American environmentalism.

When the environmental movement is criticized, it is usually this group of established organizations that is meant. They range from Greenpeace on the activist end, to large, pragmatic organizations like Environmental Defense and NRDC in the middle. Then there are more conservative groups like the Izaak Walton League, a fishing and hunting organization, with a majority of Republican and independent members. (The "Ikes," as they are called, nearly withdrew over having to associate with Greenpeace.) The Green Group also contains two Native American environmental organizations, science and public health groups like the Union of Concerned Scientists (UCS) and PSR, the National Religious Partnership for the Environment (NRPE), the Ralph Nader–founded activist consumer organization, U.S. Public Interest Research Group (US PIRG), as well as obvious and traditional environmental groups like the Sierra Club, the National Audubon Society, the Wilderness Society, and the League of Conservation Voters.

PSR lies on the liberal, activist end of this spectrum. Because it is a public health group that also deals with nukes, war, and violence prevention, and is small compared with the big enchiladas like the National Wildlife Federation (NWF) with 2 million members and a budget in the range of $45 million, I have never felt the sting of criticism toward big environmental groups as keenly as some colleagues. But as someone who came to the group from the smaller, often hard-pressed, and bitterly divided peace movement, I have been impressed by the sophisticated strategies and organizational acumen of the leaders of America's top environmental groups. Over two decades, it is they who, in large part, have helped to build our growing consciousness of global climate change. They have kept at it even during the past seven years, in the face of an administration openly hostile to environmentalism and to action on global warming.

The harshest critics of the environmental movement have their roots in radical critiques of American society and politics from the days of the New Left. They maintain the skepticism of that period toward reformism and incrementalism. At the time, this was roughly my own view of the environmental movement and of Earth Day 1970. The radical journalist I. F. Stone, one of my early heroes, called it a "gigantic snow job." It diverted attention from the Vietnam War. Antiwar critics of environmentalism were joined by even moderate civil rights leaders like Whitney Young, president of the Urban League. He said: "the war on pollution is one that should be waged after the war on poverty is won."[2]

Typical of the modern version of this critique is one of the first assessments of the contemporary movement, *Losing Ground: American Environmentalism at the Close of the Twentieth Century,* by independent journalist Mark Dowie.[3] It sets a tone of disappointment, even anger. Dowie is clearly horrified that, in 1991, Robert Nisbet and other commentators had predicted that the environmental movement would one day be seen as one of the most significant developments of the twentieth century. Dowie found such predictions "challenging, if not preposterous." The key failing of the movement has been that it "crafted an agenda and pursued a strategy based on the authority and good faith of the Federal government." The response to the Reagan administration's attack on the environment and environmental agencies, as epitomized by Secretary of the Interior James Watt and EPA Administrator Anne Gorsuch, was "genteel and anemic." The major environmental organizations were to blame for alienating and undermining "the grassroots of their own movement." Only a fourth wave of environmentalism, formed by people "outside the Beltway" with renewed radicalism and led by people of color and the environmental justice movement, can save the day.[4]

Understandably, the civil rights movement is Dowie's prime inspiration. But he mischaracterizes it as if only militancy mattered. He rejects the views and work of moderates like Thurgood Marshall, who used litigation and lobbying, while neglecting the predominately mainstream and middle-class portions of the broader movement. These included foundations, mainstream churches, lawyers' groups, wealthy funders, and ordinary Americans of all walks of life who rejected racial segregation and bigotry. And, there is no mention of imperfect, mostly moderate political leaders like John F. Kennedy and Lyndon Johnson. Such a broad movement, including reformers and strong mainstream elements, is, according to social theorist Mary Lou Finley, essential to victory.[5] A similar cast is deeply involved today, if also little noticed, in the fight against global climate change.

If early assessments of environmentalism are angry and accusatory, more recent ones still focus on the perceived failures of the mainstream. Veteran journalist Ross Gelbspan brilliantly sketched the extent and influence of the carbon lobby and the realities of climate change in his first book, *The Heat Is On: The Climate Crisis, the Cover-up, the Prescription.* Published right before Kyoto, Gelbspan's analysis puts the blame for lack of progress squarely where it belongs—on the huge energy corporations and the handful of subsidized "skeptics" they have supported. Gelbspan concludes with an impressive refutation of the direct congressional testimony of subsidized scientific climate "skeptics" like S. Fred Singer and Pat Michaels of the University of Virginia. He then notes, correctly, that American politics and opinion can undergo rapid shifts and crystallizations. World changes, like the fall of the Berlin Wall, or the end of apartheid in South Africa, can come with dramatic suddenness.[6]

I agree. But these rapid changes, like the breakup of ice formations, are preceded by years of subtle shifts and signs that are easy to miss. In his first book on climate, Gelbspan shows early signs of irritation with the American public (and perhaps the environmental movement itself). In his view, the time to act is critically short, the results of hesitation disastrous, the evidence obvious. "News stories about the warming of the planet generally evoke an eerie silence. Several front-page stories have appeared in *The New York Times,* as well as recent cover stories in *Time, Newsweek,* and *Harper's,* and a flurry of other highly visible media sources. But to my frustration, it is clear that this is a story that people do not want to hear."[7] Gelbspan is a very good investigative journalist. But, like many constant, close readers of the news, it is hard for him to believe that a few tatters of the truth do not produce immediate outrage and social change.

By his second book, *Boiling Point: How Politicians, Big Oil and Coal, Journalists, and Activists Have Fueled the Climate Crisis—and What We Can do to Avert Disaster,* Gelbspan turns against mainstream environmentalists. I had eagerly assigned his book to my graduate class in global climate change. It is written boldly and readably. It covers the science in crisp, summary form. It describes some of the positive local and grassroots developments occurring around the country. And it mentions a few innovative national projects, like the Apollo Alliance. As in *The Heat Is On,* Gelbspan is a master at making the far-reaching calamities we are starting to see visible, dramatic, and connected to global warming. Yet, my students, environmentalists to the core, disliked the book. I was surprised. But in discussing and rereading the text with them, it became clear that, although a skilled communicator, Gelbspan had misread his audience. He assumed the problem is with *them*—not as he showed in his first book—with the polluters and their paid apologists. His preface begins: "It is an excruciating experience to watch the planet fall apart piece by piece in the face of persistent and *pathological denial* [emphasis added]." Gelbspan, typically far ahead of the times, personally covered the first historic 1972 United Nations Environment Summit in Stockholm, Sweden. It listed climate change among the dangers facing humanity. No surprise, then, that for him, by 2004, "This book is a last-gasp attempt to break through the monstrous indifference of Americans to the fact that the planet is caving in around us."[8] Desperate measures are needed. A terrible tocsin must be sounded. Those who do not respond are part of the problem. They are portrayed as "too timid to raise alarms about so nightmarish a climate threat." Groups like NRDC, the Union of Concerned Scientists, and the World Wildlife Federation "have agreed to accept what they see as a politically feasible target for 450 parts per million of carbon dioxide." It "may be politically realistic," but it would probably be "catastrophic." For Gelbspan, as with Dowie and other critics, the problem is that "the major environmental groups, moreover, are trapped in a 'Beltway' mentality that measures progress in small, incremental victories."[9] To reach these "timid" Beltway goals will, of course, require an entirely new government to legislate and implement them, along with a cut of roughly 80 percent in the production of CO_2 from every corner of the American economy. An organized appeal must be made to millions and millions of Americans who are aware of global climate change, not necessarily numb or in denial. They may think it is too late to prevent the worst. More likely, they want clarified just what they can do, personally and politically, that will make a difference. In short, they would welcome powerful, principled, yet pragmatic positions.

How Green Groups Led the Fight

It is what my students of climate politics yearned for. They were aware of and appreciative of local, grassroots efforts. But they believed that such work and solutions are not sufficient for a national and global problem. They agreed with the radical critiques of the oil and coal industries and their White House and congressional allies. But their analyses led them to conclude—instead of rejecting and ridiculing national organizations—that we need *more,* not less, organizational strength and movement clout at the federal level.

Persistence, pragmatism, and professionalism are the qualities my students saw and welcomed warmly in speakers like NRDC's Senior Scientist Dan Lashof. I have described him earlier, in 1990 at the earliest scientific meetings of the IPCC. Yet in 2006 and again in 2007 and 2008, Lashof, a PhD scientist, was still patiently explaining, in cogent terms, the latest scientific findings on climate change. He has been doing this day in and day out for more than seventeen years. He constantly talks with the media, Congress, international negotiators, citizens' groups, and students nationwide. Having spoken alongside Lashof at true grassroots events—such as one in a rural Presbyterian Church in southeast Pennsylvania that drew more than two hundred activists, local politicians, and the area member of Congress—I was aware of just how absurd the frequent charge of elitism and failing to work with the "grass roots" can be.

Jeremy Leggett, for instance, describes the principled and pragmatic tactics of Greenpeace from the earliest days of its climate campaign. After he became aware of the impact of carbon emissions on global warming, Leggett, a trained, professional oil geologist, quit his position at the Royal College of Mines. He took his talents to Greenpeace. For years he made scientific presentations to insurance companies, oil executives, and the media. He lobbied tirelessly at the endless series of international scientific sessions and worldwide negotiations that have gone on since 1990. By 1990, when IPCC climate sessions began, Greenpeace had already helped, through lobbying with other environmental groups, to get a moratorium on whaling, to make progress toward a Comprehensive Test Ban Treaty, to protect Antarctica from oil drilling, and to institute a ban on chlorofluorocarbons (CFCs), which were putting a hole in the ozone layer. "Just as much effort went into the less glamorous work of corporate and governmental lobbying as was invested in harrying whalers, blocking toxic-waste pipes, and so on," says Leggett. "My new colleagues wore suits and twinsets just as often as they did anoraks and wetsuits."[10]

It is this ability to work with and address *different* audiences in *diverse* circumstances and cultural settings that is so important to the education and organizing that has marked most mainstream environmental groups. With Greenpeace, as with NRDC, the ability to support and deploy a trained expert like Leggett or Lashof over many years has been critical. Thus, by the first World Climate Conference in Geneva in 1990 (it set the terms for the treaty negotiations that followed in Rio de Janeiro), there were, in addition to Lashof and Leggett, dozens of professional environmental staffers from NGOs. They lobbied negotiators, put out an eagerly awaited daily report to delegates and the world media called *ECO*—a newsletter still published at each climate negotiation. The first climate summit in 1990 featured a Greenpeace demonstration by their German and Swiss members who, chained together, blocked the main and side entrance to the UN's Palais de Congrès as a huge balloon floated overhead urging delegates to "Cut CO_2 Now." Inside, the first Bush administration had failed to send even the head of the EPA to the summit. The heavy lifting was left to John Knauss, a low-level official from the Commerce Department. Knauss stalled and stonewalled as best he could. But the real drama, according to Leggett, came from the pleas and demands of the so-called small island nations, places like the Solomon Islands, Trinidad and Tobago, Kiribati, Barbados—some thirty nations in all—who stood to be thoroughly destroyed by rising seas. This group had formed a historic UN coalition called AOSIS (the Association of Small Island States). It was confronting head on giant, rich nations like Japan and the United States. The interventions from AOSIS were powerful, especially that of tiny Kiribati. Its president, Ieremiah Tabai, had met with Leggett and a Samoan Greenpeace member in Kiribati prior to the summit. It was Tabai who helped hold off attempts by the United States, Saudi Arabia, and the carbon club to negate the seriousness and urgency of climate change. The world press (less so in the United States) took notice; global climate change was given a dramatic, human face. But as it turns out, AOSIS, formed at Geneva, had its founding principles drafted by two middle-class professionals, environmental lawyers James Cameron and Phillipe Sands. They incorporated the "precautionary principle" into the AOSIS language. They created a radical tool still beloved by grassroots activists. It embodies the public health approach. It says that further delay or inaction is uncalled for when there is sufficient, plausible science—even if not absolutely conclusive—to show significant human harm.[11]

Such combinations of tactics—lobbying, diplomatic interventions, legal innovations, demonstrations, and work with both U.S. and world media—continued at every session, whether of scientists, ministers (cabinet members), or actual treaty negotiators. At these sessions, environmentalists also formed working alliances with American and world political figures. In 1991, after speaking to the Japanese parliament, Jeremy Leggett made his first presentation at a nearby meeting in Tokyo. It was arranged by a Greenpeace member, Yasuko Matsumoto, who had been lobbying the legislature in Japan. The group Leggett addressed about the potentially surprising and catastrophic consequences of climate change was GLOBE (Global Legislators Organisation for a Balanced Environment). Its president was none other than U.S. Senator Al Gore. GLOBE had been formed on the model of earlier anti-nuclear groups of parliamentarians initiated in the United States. It had already been meeting with U.S. environmental groups lobbying on Capitol Hill and at the United Nations. These American groups (most were not truly international like Greenpeace) had formed the U.S. Climate Action Network (USCAN), part of the broader international formation, ICAN. Thus, from the earliest days of sustained movement work on climate, the environmental movement and its mainstream U.S. members used a variety of tactics: nonviolent action, public and grassroots education and organizing, lobbying on Capitol Hill, and monitoring, lobbying and intervening at UN climate sessions. Throughout, they met with both U.S. and world media, provided expert and scientific information, diplomatic and legislative strategies, and news and rumors that other parties to this complex process simply did not have. Then they communicated the same information and action suggestions to their own growing grassroots networks.[12]

During 1991, just before the Rio Conference that created the Framework Convention on Global Climate Change, these dogged and rather undramatic reform efforts began to pay off. By December 1991, Senator Al Gore flew directly to the negotiating session in Geneva. He met with environmental lobbyists and attempted to push the U.S. delegation to commit itself to specific targets for cutting carbon emissions. As the session ended, Michael Oppenheimer, a world-class astrophysicist who had worked on the scientific assessments, then spoke for Environmental Defense and thirty other environmental organizations at a packed press conference. Oppenheimer, whose book on climate change came out that year, is widely respected in both the scientific and environmental communities. Speaking with unusual emotion, he derided those who claimed there was no scientific consensus on the dangers of climate change. He denounced the U.S. government's failure to take serious action, as well as the Saudis and Kuwaitis, whose "commitment to oil outweighs their commitment to humanity."[13] Such boldness and moral passion is not uncommon among mainstream leaders. Oppenheimer, who now holds a prestigious chair at Princeton University, still remains active with activist climate campaigns. He has spoken, like Lashof and others, in venues large and small, far and wide. If this be elitism, then we need more of it.

Given the heavy corporate opposition to any attempts to get mandatory reductions of greenhouse gases, there were also early and shrewd efforts by environmental organizations to muster economic arguments and find business allies. These began with attempts to splinter and block the efforts of the Global Climate Coalition. The cracks in this powerful industry lobby, as with climate change itself, were at first small and subtle. The Edison Electric Institute, representing American utilities, was the first to go. In July 1996, when Undersecretary of State Tim Wirth announced in Geneva at COP2, the second Conference of the Parties that preceded Kyoto, that the U.S. government would finally support mandatory limits on CO_2, this new position—a clear result of lobbying by environmental groups and internal battles in the administration—was immediately attacked by most energy corporations. The Global Climate Coalition had already sent a letter hostile to formal mandates to the White House before the negotiations. It was signed by 119 corporate CEOs from Amoco, ARCO, Chevron, Chrysler, Exxon, and a host of other oil and coal groups. But Charles Lindermann, then the representative of the Edison Electric Institute, refused to endorse the GCC statement. "We sell electricity," he said in a press conference, "we do not care where it comes from. The power lines do not know the difference if it comes from coal, a windmill or a solar cell. We are with the future, not the past."[14]

Then, the United Nations Environment Programme's insurance initiative, citing losses to insurers called for "early, substantial reductions in greenhouse-gas emissions." Finally, it was in Geneva that the Business Council for a Sustainable Energy Future first appeared. Environmental groups welcomed and worked closely with it. Although brand new and far smaller than the Global Climate Coalition, the Council represented more than twenty companies and organizations, mostly from the energy efficiency and renewable sectors like the American Wind Energy Association. Among them, as Leggett points out, "was Enron, America's largest gas company. As the *Oil and Gas Journal* had feared, civil war had broken out."[15] Soon, oil companies like Shell and BP and auto giants like General Motors and Ford began to sound concerned about climate change. They did not withdraw from the GCC right away. But signs of cracks and melting in the huge industry block slowly widened.

Though their numbers were relatively few, the efforts of such climate change pioneers and environmental organizations slowly began to have a political effect. By the spring of 1992, the recalcitrant White House of George H. W. Bush started to feel pressure. In April, Rep. Henry Waxman (D-CA), chair of the House Subcommittee on Health and Environment, who had been meeting with environmental groups, introduced a bipartisan Global Climate Protection Act. It soon had nearly 150 cosponsors. It called simply for a U.S. commitment to freeze its level of CO_2 emissions. Yet the White House Council on Environmental Quality (CEQ) wrote to every member of Congress, opposing the bill. CEQ denounced Waxman and his assertion that the United States was the central obstacle to calling for stabilization of CO_2. Waxman responded by citing numerous officials from the European Commission, Japan, Canada, and the Netherlands who agreed with him. Then he called on President Bush to reconsider his position. This congressional attention and criticism of the U.S. stance next caught the attention of one Bill Clinton. Clinton was just emerging as the frontrunner for the Democratic presidential nomination. On Earth Day, April 22, 1992, he attacked the first Bush administration and its ties to big polluters, saying that "our addiction to fossil fuels is wrapping the earth in a deadly shroud of greenhouse gases."[16]

Given such sophisticated organizing by environmentalists, by June 1992, at the Earth Summit in Rio de Janeiro—despite the best efforts of the carbon lobby and the oil-friendly Bush administration to avoid it—the world signed its first global climate change treaty. It called for the reduction of CO_2 and other greenhouse gases. At that summit, with some four thousand members of the media in attendance, along with thousands of delegates, Al Gore, who had just published *Earth in the Balance,* attended in person. He upstaged President Bush, who had at first refused to go. Dozens of environmental leaders from the established groups and from the Climate Action Network—all now seasoned negotiators, publicists, and organizers on a critical world environmental stage—roamed the halls, familiar and highly effective, if unheralded, advocates. The Rio Treaty fell short of mandatory limits on carbon emissions; it reflected the foot-dragging of the Bush administration and the relentless work of the carbon lobby. Nevertheless, mainstream environmentalists and a few of their allies in Congress had made history while few others paid attention.[17] Mandatory limits on CO_2 might yet be needed to push businesses into innovations. But a signal had been sent to world energy markets. And the scene was set for the Clinton administration.

Environmental Organizations and the Clinton Years

It would take five more years until the Kyoto treaty, with its mandatory cuts in greenhouse gas emissions, was signed. Nearly four of them were under President Bill Clinton and Vice President Al Gore. Here again, mainstream environmentalists carried out skillful campaigns against great odds. It is now rarely recalled that Bill Clinton was elected as a minority president. He received only 43 percent of the vote in 1992. Without the role of Ross Perot as spoiler with 19 percent of the vote, George H. W. Bush would have been reelected. And by 1992, the U.S. Senate fell into antienvironmentalist Republican control, as did the entire Congress with the Gingrich "revolution" of 1994.[18] Nevertheless, critics assume that with Al Gore as vice president, far more should have been achieved. Once again, the claim is that mainstream environmental leaders sold out the grassroots. They were content with mere access and breakfasts with power brokers in Washington. With Gore and his former aides Carol Browner at EPA and Katie McGinty at the White House Office of Environmental Quality, and Tim Wirth at the State Department in a newly created environmental post called the Under Secretary of State for Global Affairs, environmental groups did indeed have real access. But they had to push hard to make it meaningful. They remained, in the words of John Adams, the highly respected leader of NRDC, "outsiders."[19] It is fair to note that Clinton did little for the environment in his first year. Gore, then vice president, not an independent senator and *Earth in the Balance* author, did not help much initially, either. He infuriated activists by approving the start-up of the East Liverpool, Ohio, incinerator that he had publicly opposed when not in power. The Clinton-proposed free trade NAFTA Treaty, despite some minimal environmental and labor protections, was supported by most mainstream environmental leaders, further riling the grass roots. But contrary to the views of critics, within a year of the jubilant Inaugural, Green Group leaders had strongly expressed their disappointment and displeasure with Bill Clinton's first year. It was this displeasure, not toadying, that led to a breakfast meeting with Al Gore, who angrily walked out before ultimately setting up regular meetings.[20] But when Newt Gingrich and other antienvironmental conservatives swept into power in 1994, it was again national environmental groups that took the heat.

I joined the Green Group at this point. I got a very different picture from that often painted by outside critics. First, the nature of the group was already far from its origins in the 1980s as the smaller and more elite Group of Ten. The founders and dominant personalities, Jay Hair of the National Wildlife Federation and Peter Berle of Audubon, were gone. I had met both briefly and found them cool and somewhat dismissive. At least on the surface, they met the stereotype of self-confident WASPs. Their style was top-down and buttoned-down. But the idea that they were out of touch with the grass roots and simply sold out inside the Beltway reflects the perennial bias of critics. They bemoan the power of organized lobbyists from industry or the military-industrial complex in Washington. But then they denounce those who actually go wingtip to wingtip with the special interests who would otherwise roam the marbled halls of power unchallenged.

For instance, at PSR, I had hired the National Wildlife Federation's former top lobbyist, Sharon Newsome. She had come from grassroots organizing in Alaska and been so successful that she rapidly moved up in environmental circles. A fierce competitor and critic of polluters, Newsome could alternately wield a sharp tongue and temper, or smooth sophistication. She worried, correctly at times, that certain legislation pushed by environmentalists would ultimately backfire, causing a backlash of revolt among ordinary, working-class Americans. She often fought with self-appointed radicals, but only because her own grassroots background—which she never touted—allowed her to understand how average Americans might react. For her, this was the case with mandatory tailpipe emissions inspections to control or reduce NOx, SOx, and other emissions from burning gasoline. The regulation makes sense and has proved an important public health measure. But the burden has fallen on Americans with older cars that require expensive repairs and tune-ups. More affluent drivers simply bought new ones that proved to be bigger and less efficient and that use more gas. The point here is that there has often been heated debate within and among large environmental groups. They frequently have professional staff with long backgrounds in grassroots organizing and memberships composed of equally committed Americans. Some, like NWF, with its 2 million members, reflect the views of a constituency far more typical of the majority of Americans than smaller, self-proclaimed "grassroots" groups. The same has been true of the Audubon Society. Like NWF, it has an extensive grassroots network of chapters and activists in communities throughout the United States. If by *grassroots,* one means active members around the country engaged in educational and political activities in favor of environmental protection, it is absurd to argue that mainstream groups like NWF and Audubon are elite and out of touch simply because they are large, mostly middle-class, and more conservative than local radical groups, Greenpeace, or the PIRGs.

Within the Green Group, the grassroots are also represented by far more liberal groups. The US PIRGs—whose longtime leader, Gene Karpinski, has been chair of the Green Group and now heads the League of Conservation Voters—started with Ralph Nader. Karpinski consistently fought for tough stands in meetings with the Clinton administration. The US PIRGs are backed up by a decentralized, grassroots network of state PIRGs and by a campus network with paid staff organizing at some sixty-three U.S. college campuses.[21] The PIRGs have been willing to litigate, lobby, educate, and demonstrate, both at federal and state levels. And they have been able and willing to make the connections between families, pocketbooks, and health in local communities. For example, they simultaneously attack coal-fired utilities as stimulating global warming, polluting the air, and producing asthma.

That's why, in Seattle, Washington, after Kyoto, I stood along the waterfront beside a large, inflated, and somewhat garish plastic PSR "global thermometer" designed for TV cameras. Next to me, lugged in by Washington State PIRG, was a huge, inflated plastic model of a coal-fired utility. Both PSR and the PIRGs have local offices, local staff, and grassroots activists and members throughout the Puget Sound area. Back in D.C., when meeting with Carol Browner or other Clinton officials, Gene Karpinski and I, or our legislative directors, might have been polite. But our message from the grassroots was unmistakable. The same is true of the Sierra Club, which has numerous local branches, three-quarters of a million members, and a grassroots governance system that has come close to forcing reactionary, racist immigration policies on the Club. Like US PIRG, the Sierra Club also maintains a network of student Sierra Clubs, with paid staff on numerous campuses.

Even Environmental Defense (ED), often maligned by activists and radicals, was at times strongly critical of the Clinton administration. One of its board members, the late David Rall, was founder of the National Toxicology Program, and a pioneer in tracing the effects of chemicals on humans. In the 1990s, he joined PSR as well. The stereotype of the kindly Marcus Welby physician (he looked the part), Rall was a deeply committed and fierce advocate for protecting human health. He was relentless in pushing to improve the standards at EPA as physicians and scientists learned more and more about the effects of even tiny doses of certain chemicals on the human endocrine system. Along with Rall, ED senior scientist Peter DeFur, and others, PSR and ED jointly produced a major report on dioxin critical of the failure of the EPA to implement new standards.[22] Dioxin, one of the most dangerous chemicals known, is produced as a by-product of burning or incinerating any substance with chlorine in it. The ED/PSR report led to "insider" meetings at EPA where PSR and ED were seriously listened to. But it was also sent nationwide to local media, to PSR chapters, to other local groups, and released in prepared video releases. These led to some 10 million Americans being exposed, not to dioxin, but to highly critical news about our government's failure to adequately protect American families.

There is little doubt that the mainstream movement at first put too much faith in litigation and legislation alone. But as the need for new allies, new tactics, and new issues emerged in the late Bush and early Clinton years, the mainstream movement changed fairly dramatically. The old characterizations that still linger were increasingly poor sketches.

The Green Group Expands Its Horizons

The reaction to the Gingrich revolution of 1994—when conservative, anti-environmental Republicans far removed from their Teddy Roosevelt roots took over the House of Representatives—was neither tepid nor timid. The initial meetings of the Green Group after the election were indeed dispirited, almost despairing. But after a requisite period of shock, the Green Group mounted its first unified, grassroots response with a platform Pledge of Environmental Values. It engaged millions of Americans in learning that the Contract with America was what environmentalists dubbed the Contract *on* America. The strategy was to combat the notion that the new House had the broad public interest in mind, to point out the antienvironmental (and hence anti-neighborhood, anti-health, and anti–family values) positions that went with it, and to mount counterattacks locally and in Washington.

The Green Group also expanded in the '90s to include new entities that did not have grassroots affiliates, members, or governance structures. These new groups, like the Environmental Working Group (EWG) and the National Environmental Trust (NET) added pizzazz to the older organizations. Unhindered by the complex demands of membership and grassroots affiliates, NET and EWG were far more nimble and able to mobilize quickly around new threats to the environment. They used tactics and techniques often better suited to the information age blossoming in the '90s than did older membership groups.

The Environmental Working Group, for instance, is headed by media-savvy Ken Cook. He and his longtime associate, Richard Wiles, go back to the early '80s when they were young Turks desperately trying to fight off Reaganism. Together they had pioneered the first truly sophisticated use of computers by environmental groups. They regularly mined federal data from the EPA, the Toxics Release Inventory, and other sources. With their finds, they created hard-hitting, media-friendly reports and amazing interactive Web sites that mapped the effects of environmental toxins. PSR first teamed with EWG on a report called *Tap Water Blues*. It traced the impact of the widely used agricultural pesticide, atrazine, and its human health effects after it got into drinking water throughout the Midwest Corn Belt.[23] The report was widely cited, got serious media coverage, and was an effective tool in the lobbying to make Americans' drinking water not only cleaner, but *healthier*. EWG has now followed this pattern for years. As Web technology has improved, EWG has been able to mount campaigns on behalf of more obscure, yet critical, environmental issues like nuclear waste.

Cook and I took the lead in some key lobbying meetings against the Yucca Mountain nuclear waste repository. In one, we met with then head of the Senate Environment and Public Works Committee, Jim Jeffords. A friendly, if reserved, Yankee moderate, he had left the Republican Party to become an Independent. I talked about the radiation risks of transporting nuclear waste across the nation to Yucca, Nevada. These included exposures to the drivers of rigs carrying huge concrete (but not lead-lined) casks. Cook then described the nationwide transportation necessary and the outpouring of grassroots response to a special EWG Web site. It had been created to allow any American to click on a map and find out through which towns nuclear wastes would travel to get to Yucca. As a result, hundreds of thousands of Americans had already sent e-mails of protest to Congress. The Yucca campaign also included a traveling road show in which the PIRGs literally trucked a mock nuclear waste cask around the country. At the end of the tour, they parked the rig in front of the Capitol for a rally with short speeches by the truckers, Gene Karpinski, me, and other Green Group activists. Such grassroots action was matched with a powerful set of television spots that PSR ran throughout the campaign on the dangers of nuclear wastes and Yucca. What had been an obscure, nuclear, and NIMBY issue became a major national environmental battle. Ultimately, the campaign defeated Yucca Mountain; it was a major victory against powerful, entrenched interests.

There are many other grassroots national organizations in the Green Group with local affiliates, local volunteers, and a network of activists to join in on nationally orchestrated lobbying campaigns. They include American Rivers, Scenic America, Rails-to-Trails Conservancy, the Izaak Walton League, the National Parks Conservation Association, the Trust for Public Lands, and others. All told, perhaps 10 million or more Americans belong to Green Group organizations. Many of them are active in local projects, education, and lobbying. They are increasingly diverse and have significant numbers of members, perhaps millions, who are rural, working-class, and surely not Ivy League lawyers.

Perhaps the best example of the failure of critics to recognize significant changes is the Green Group's inclusion of the leading religious environmental organization into its ranks (along with Native Americans, organized state groups, and others). The media now notes with some surprise that religious groups, especially evangelicals, seem suddenly interested in global climate change. But a number of environmental leaders were aware that as early as December 1991, in Geneva, the World Council of Churches had issued a major pronouncement on global climate change. It said that "concern for the protection of Creation is growing rapidly in the churches . . . with global warming being a high priority."[24] By 1994, the World Council of Churches had issued a report, *Accelerated Climate Change: Sign of Peril, Test of Faith* that was cited at the UN. It warned that global warming was a threat to life and "an issue of justice."[25] That same year, religious leaders and scientists like Carl Sagan were organized by the Union of Concerned Scientists and others into a major environmental summit. Climate was a key issue. By 1996, not long after its formation, the National Religious Partnership for the Environment (NRPE), headed by Paul Gorman, was seated with the Green Group. NRPE, composed of mainstream Christian and Jewish bodies (and to a lesser degree Muslims and other faiths), attended international negotiations, White House lobbying sessions, and organized at the grassroots.[26] The National Council of Churches also developed a global climate change program in earnest around this time, leading and coordinating efforts across the United States through state Councils of Churches and local congregations, as well as Washington lobbying.

Building on their own engagement with politics and lobbying that had emerged with the Vietnam War and the anti-nuclear movements, mainline religious environmental leaders in the '90s brought a wealth of grassroots energy, political savvy, and a fresh, large constituency to environmental and climate battles. Advocating on behalf of many millions of mainstream religious citizens concerned with the environment, Paul Gorman got serious attention in Clinton circles. He frequently spoke with a prophetic, though soft and somewhat whispery, voice. He was also backed by others who, at his urging, had attended a series of spiritual retreats for the Green Group. These were funded and arranged by the Nathan Cummings Foundation, a liberal Jewish philanthropic organization. The environmental leaders at these sessions sought to deepen their own sense of vision, leadership, and ethical engagement for the obviously new, major challenges before them. But they also gained understanding and appreciation for religious perspectives ranging from Zen Buddhism and indigenous Hawaiian practices to the influence of St. Francis of Assisi, the Roman Catholic patron saint of animals and nature, to other forms of environmental Christianity and Judaism that were presented and studied. It was also around this time that Carl Pope, the head of the Sierra Club, gave a widely circulated address apologizing for environmentalists' mischaracterization and overemphasis on the negative impacts of Judeo-Christian thought on the environment ("Go forth and multiply," "I give you dominion," and so forth).[27]

The Struggle for the Kyoto Treaty

As we in the Green Group pushed both for environmental health measures and for the strongest possible targets and timelines for what would become the Kyoto treaty, Gorman was able to appeal directly to the moral sensibilities of Gore and others. He spoke for a newer, broader environmentalism with a positive vision. It also created a useful kind of good cop, bad cop effect alongside more aggressive types like NET's Phil Clapp. Clapp, drawn to effective media stunts with a tinge of ridicule, sometimes drew ire, either muted or overt, especially from Gore. As we collectively pressed to get the vice president to attend Kyoto as he had Rio and other negotiating sessions (it looked as if he would not attend), Clapp sent out copies of *Earth in the Balance.* It went to all the major news outlets in the nation. Inside, heavily underlined, were the lines where Gore had scorned George H. W. Bush for at first being afraid to attend the Rio Conference.

Environmentalists had helped pressure the first Bush administration into the Rio Framework Convention on Climate Change of 1992, even if it was voluntary. Pushing hard to achieve a follow-on treaty with teeth seemed the obvious thing to do. And so, in December 1997, along with other environmental leaders, I headed to Kyoto, Japan. PSR had helped to create, with the Harvard Center for Health and the Global Environment, a national statement by hundreds of physicians and health professionals, published as an ad in the *New York Times.* It reiterated the validity of the science linking climate change to human activity, stated the threat to human health from disrupted climate systems, and called for a strong international treaty.

Each day, the scene outside the Kyoto Convention Center belied the tense, often frenzied or angry negotiations and lobbying going on inside. There were brilliant, sunny, late autumn skies with cool temperatures. A small lake nestled behind the Center with pine trees, birches, chickadees—a scene, but for strolling Japanese families, that might have been from *On*

Golden Pond. In front of the entrance to the sprawling concrete modern building stood, amusingly, ice sculptures of penguins gently melting. Those of us not official delegates spent our time on a huge convention floor with perhaps four thousand members of the international media and NGOs. Jammed in with us were the lobbyists, spinners, and climate skeptics representing the carbon lobby—the Global Climate Coalition, that Orwellian-named collection of about one hundred of the leading oil, coal, and gas corporations.

I recall vividly the moment when the treaty was completed. It followed days of sleeping under chairs, of prowling the desks of reporters with tidbits of news, press releases, or gossip, and plunging headlong into scrums of pro- and anti-treaty proponents, fuzzy media mikes, and hand-held recorders each time a recognizable delegate entered this bedlam. The final treaty results had been leaked before they blared forth on big video screens throughout the hall. A Kyoto Protocol calling for 7 percent reductions in U.S. carbon emissions below 1990 levels by the year 2012 had been achieved.[28]

What I remember at the announcement is the exhausted exhilaration of American environmentalists, friends and colleagues from the Sierra Club, Environmental Defense, the NRDC, the National Environmental Trust, the World Wildlife Fund, the Union of Concerned Scientists, Greenpeace, and more. But mostly I can still see the defeated, slumped, and frustrated postures of the slick and once mighty representatives of the Global Climate Coalition. They stumbled out into the cool fall Kyoto air, past the penguins, and onto buses, cars, and taxis that carried them off, home to the United States, in ignominious defeat. Clearly Kyoto was a significant achievement. It was seen by the carbon lobby as a serious setback. It is why they soon mounted a major counterattack for which, frankly, the major environmental groups were ill-prepared and ill-suited.

When we returned to the States, the framing of the event for the media, participants, policymakers, and the public was already being set. The treaty was unfair to the United States. The mandated U.S. cuts in carbon emissions—based on mere theory—would hurt jobs. And the subtext read: isn't this really just the work of a gaggle of elitist international scientists, bureaucrats, and environmentalists in league with Al Gore (Ozone Man) and Bill and Hillary Clinton? Some parts of the charges implicit in the questions raised had elements of truth in them. Environmentalists, perhaps, had it too easy with champions—however beleaguered and cautious—in the White House, State Department, and EPA. In plain terms, before Kyoto the environmental movement had not built much public understanding or grassroots support for action. Without the election of the minority Clinton government in 1992, Kyoto would not have been possible. But with it, the focus for the movement became pushing the Clinton administration directly and monitoring and lobbying the complex international negotiation sessions and scientific debates that preceded Kyoto.

The Green Group had publicly and privately pressured the administration to back a strong Kyoto treaty and other environmental measures. But more typically, the Clinton-Gore White House looked to major environmental organizations to carry out media and organizing efforts that they could not, to lobby the Congress, and to push against the opponents of Kyoto within the Cabinet itself. That's how I met Larry Summers, then secretary of the treasury. Summers believed, as was conventional wisdom at the time, that mandating deep cuts in carbon emissions would cost a significant portion of the federal budget, harm American competitiveness, and cost jobs. We environmental leaders came armed with economic reports and analyses from Robert Reppetto of the World Resources Institute, indicating that the American economy would actually improve. There would be new efficiencies, technologies, jobs, and other innovations. These would all come from squeezing out the outmoded oil economy. Summers took them about as seriously as he did the Faculty of Arts and Sciences or women scientists before he left the presidency of Harvard after the shortest tenure since the 1830s. But others in the Clinton Cabinet were quite engaged, especially Carol Browner of EPA. She had Gore's ear and was committed, smart, and persuasive. Undersecretary of State Tim Wirth, a former senator who remained close to Clapp and NET, was also helpful. In fact, in most cases where former environmental staffers or officials were in positions of influence in the Clinton administration, they welcomed environmental groups and their former colleagues. They did not use them for compromising Clintonian ends, as critics have claimed. Administration officials looked to the big green groups to get reinforcements for arguments they were making *inside* the administration.

But what has been a serious problem for the Green Group and other environmental organizations since the heyday of environmental successes in the 1970s is the difficulty of portraying newer, less visible environmental threats to the American public. Most of the traditional environmental groups were ill-equipped by mission, history, staffing, and membership demands to move quickly into the newer areas of environmental health and toxic pollution, let alone worldwide threats like global warming. Nor has funding been as munificent as small local groups or journalistic and academic critics might imagine. Only in 1998, after Kyoto, did one of the largest liberal foundations, the John D. and Catherine T. MacArthur Foundation, give three-year grants totaling a million dollars to each of four active climate groups: UCS, WWF, ED, and NRDC. A smaller set of grants went to groups like PSR, which received seventy-five thousand dollars a year.

I recall vividly being asked sternly by the program officer at MacArthur whether PSR was committed to the long-haul on climate change. The foundation was only interested in groups that were. I assured my grant officer that ours was indeed a firm pledge. Besides, we had been fighting nuclear weapons since 1961. We knew something about the value of commitment. PSR is still working on climate change. Tragically, the MacArthur Foundation is not. It stopped funding the entire climate area after the three years were over (just as George Bush arrived in 2001). Later, in 2004, MacArthur also backed out of the Energy Foundation it had helped to create and fund. Meanwhile, just as the Bush administration was installed and as the United States was attacked on 9/11, other major funders of climate work disappeared. The W. Alton Jones Foundation split up over a family dispute and ceased operation. The Turner Foundation stopped giving grants and went into hibernation as Ted Turner's AOL stock plummeted from $52 to $9 per share. And, in the world of direct mail, much derided by purist critics, pragmatic Americans responded with their hard-earned dollars far more generously to appeals focused on more immediate problems.

Until very recently, the major environmental groups actively involved in climate work over time since the '90s could be counted on one's fingers: ED, NRDC, Sierra Club, Greenpeace, the National Environmental Trust, World Resources Institute, the World Wildlife Fund, and more specialized groups like the National Religious Partnership for the Environment, the Union of Concerned Scientists, and Physicians for Social Responsibility. A few others, like the PIRGs, have been closely involved in related clean air campaigns but, until recently, not with climate change per se.

Thus, in the Clinton years, only a few relatively modestly funded national organizations pushed against and pulled with the Clinton administration to get climate change onto the map and to deliver the Kyoto treaty to Americans and to the world. They did this while simultaneously protecting clean air standards, forests and wilderness areas, and while creating a Children's Health office within the EPA to develop a tenfold safety factor for kids' exposure to pesticides. Perhaps the most balanced assessment of the main environmental movement comes from Phil Shabecoff, a former *New York Times* reporter. He has followed environmental issues for years and founded the environmental news service GreenWire. Opposed to the idea that the main environment movement wilted or sold out during the Reagan years and the growing counterattack from industry and the White House, Shabecoff says: "a number of environmental organizations, chiefly those operating at a national level, sought to develop new skills and tactics. They developed expertise in economic and political analysis, adopted more aggressive media and public outreach strategies, paid at least lip service to the disproportionate ecological ills heaped on the nation's poor and minorities, and looked for ways to achieve their goals without the assistance of the now less-than-sympathetic Congress and courts."[29] Much of what was achieved (or saved) was not related to climate change; it only became a visible issue in 1988. Recall that it was still perceived, until well after 2001 and the Third IPCC Assessment, as a somewhat muddled scientific controversy. But, concludes Shabecoff: "it was clear that environmentalism had wrought profound changes in American life—to its landscape, its institutions and its people. That our air is somewhat clearer and more breathable, that our water is somewhat cleaner and more drinkable in many places, that we are not buried in garbage, that some abandoned toxic waste sites have been cleansed, that some of our wild lands have been preserved from development and some of our threatened wildlife has been protected constitute almost miraculous achievements in the face of the economic juggernaut."[30] Unlike those who call for an entirely new movement or for purely radical grassroots action, Shabecoff sees the need for a larger, even more sophisticated, more political movement. No longer can environmentalists count on broad but politically disengaged public sympathy. What is needed is political power, "power that will put them in a position not just to ask or argue or beg but to demand that government do what is necessary."[31]

This assessment is key to understanding recent changes in the well-organized national environmental movement. Working to get Congress and the next White House in the hands of pro-environmental officials is essential. Radical environmentalists, some grassroots activists, Greens, and academic and journalistic critics may have scorned the delicate dance between mainstream "enviros" and the Clintonistas. But, in 2008, it should be obvious that the structures and achievements of environmentalism from the Clinton years need to be built upon, not weakened or derided.

6

The New Climate Movement

An even bigger climate movement is still needed. The good news is that it has been under way for some time. It has the opportunity to unify disparate constituencies, organizations, and grassroots operations. But the established organizations long at the center of climate change efforts should remain there. Some of the most important developments in environmental groups and new climate change mobilizations include alliances with and influence on business, work with religious communities, a renewed emphasis on state and local organizing and college campuses, more sophisticated nationwide lobbying and media campaigns, and plain old electoral politics. The environmental movement has definitely moved beyond its traditional base. It includes segments of society—like American business—that it once neglected, even scorned.

Business and Action on Climate Change

In 2000, the Global Climate Coalition, long the corporate nemesis of major environmental organizations, closed its doors.[1] The Green Group did not. But the arrival of the Bush administration provided relief to die-hard climate polluters like ExxonMobil. Nevertheless, by that time, the pendulum within business was already beginning to swing toward clean energy. There were increasing signs that the splits in the carbon club were becoming chasms. Even the *Wall Street Journal,* long a cheerleader for climate skepticism, ran a special issue called "The Race to Profit from Global Warming."[2] When chief executives of hundreds of global corporations gathered at the World Economic Forum in Davos, Switzerland, in February 2000, they voted for the first time that the number one worry for business was global warming.[3] Today, entire portions of the energy industry have broken off from the carbon lobby mainland, in the same way that the Larsen B ice shelf in Antarctica fell into the sea. Efforts to involve business in solving climate change that might have been scoffed at by some hard-core environmentalists during the Clinton years have begun to bear fruit. Following her years as deputy undersecretary of state working with Tim Wirth, Eileen Claussen initiated the Pew Center for Climate Change. Members agree to cut carbon emissions and work toward sustainability. Beginning with corporations like Boeing, Enron, Lockheed Martin, 3M, and United Technologies, the Pew group has now expanded to forty-three members.[4] Financial and insurance firms have responded as well. After making the rounds presenting to and lobbying insurance companies for years, Jeremy Leggett and others have seen big players like Swiss Re come aboard. Citigroup, Chase, Goldman Sachs, and other banking and investment firms have begun to back green technologies and to speak out.[5]

Joseph Romm is a former assistant secretary of energy from the Clinton administration. He directed its Office of Energy Efficiency and Renewable Energy. Like Eileen Claussen, he has been hard at work on the outside during the Bush years. It was his office that issued an 1997 energy report that estimated that carbon emissions could be cut by U.S. business without harm to the economy, as was widely claimed at the time.[6] Now the head of the nonprofit Center for Energy and Climate Solutions, Romm's book, *Cool Companies: How the Best Businesses Boost Profits and Productivity by Cutting Greenhouse Gas Emissions,* describes how more than fifty major corporations like DuPont, Xerox, Toyota, Compaq, and 3M were among the first to improve their profitability and reduce emissions by greater energy efficiency in manufacturing. Unheralded changes in operations including motors and motor systems, enhanced office and building design, and switches to "cooler" power such as cogeneration, solar, wind, and geothermal boosted bottom lines while cutting carbon emissions.

Given the criticisms from hard-core greens for even lobbying the federal government, it is ironic that Romm thanks energy efficiency guru Amory Lovins of the Rocky Mountain Institute—a pioneering, but outsider voice on energy when Beth Parke and I first carried him in the 1980s on *Consider the Alternatives,* our nationally syndicated radio program—for recommending him to the Department of Energy for his first job there[7] Romm also worked with and thanks Dan Reicher, special assistant to the DOE secretary. Reicher, an NRDC veteran, was mocked by harsh critics like

Mark Dowie in the 1990s for seeking not only access to power, but for actually working *inside* the Energy Department.[8] Reicher, it should be noted, in league with public-interest groups like NRDC, Physicians for Social Responsibility, and others, was the point man in helping to move then neophyte Energy Secretary Hazel O'Leary to push for a Comprehensive Nuclear Test Ban Treaty.[9] She in turn helped persuade others in the cabinet, and ultimately Bill Clinton, to sign this critical treaty.

So it is with American business. These are huge institutions, difficult to change overnight. They, too, have their visionaries, ethical professionals, and individuals who care as deeply about the environment or peace, their families and communities, as do their critics. The trick is to find and communicate with them. Take Shell Oil, for example. I've cheered on Greenpeace activists climbing walls to unfurl protest banners as Shell was complicit in the murder of dissidents in oil-rich Nigeria. And I have boycotted Shell, ExxonMobil, and others over oil spills, attacks on climate science, and more. But these are tactics designed to shine the light of media publicity on huge companies. They are never sufficient in themselves. Insider tactics also play crucial roles. For Royal Dutch Shell, the two approaches have had a kind of synergistic effect. Inside the company, as early as the 1970s, Shell's planning group, led by Pierre Wack, forecast both the oil-price boom of 1973 and the bust of 1986. Sounding a bit like George Lakoff, Wack has said: "novel information, outside the span of managerial expectations, may not penetrate the core of decision makers' minds, where possible futures are rehearsed and judgment exercised."[10] Thus, Wack wrote a special set of future scenarios designed to create a paradigm shift at Shell. Business as usual would not work in a world where oil demand would inevitably drop. Later, while Shell was still part of the Global Climate Coalition, Ged Davis, head of scenario planning for Shell International, wrote a company paper called "Global Warming: The Role of Energy-efficient Technologies." By 1995, these ideas about renewables became commonplace in the speeches of chairman and CEO of Shell UK, Chris Fay, who said: "There is clearly a limit to fossil fuel." He added that there would be a gap between supply and demand that could be filled for a time by hydroelectric and nuclear power, "but far more important will be the contribution of alternative, renewable energy supplies." Insights like these have driven Shell into the solar energy business and the production of fuel cells that will allow the decentralized use of power. Under Shell's Sustained Growth Scenario such projects will be the most rapidly growing form of commercial energy after 2020. Annual sales by 2030 could exceed $100 billion.[11]

To be sure, Shell was also reacting to public protests over Nigeria and its plans to sink a giant oil storage tank called Brent Spar to the floor of the North Sea. Landings on Brent Spar and a European boycott organized by Greenpeace drove Shell to reconsider and to recycle the giant tank. The company was also moved by criticisms of its participation in the Global Climate Change Coalition and by the cuts required under the Kyoto Protocol.[12] But it is the combination of tactics from all the dreary, exhausting lobbying at climate negotiations by large environmental organizations, to 1970s energy reports trumpeting renewables and efficiency by Amory Lovins, to well-planned public actions and boycotts by a major group like Greenpeace that have, mostly unnoticed, made the difference.

In the United States, auto giant Toyota, now well-known for its popular hybrid car, the Prius, also instituted energy efficiency at its huge production facilities. Toyota Auto Body of California (TABC) manufactures and paints the rear deck of Toyota pickup trucks in Long Beach. TABC consumed 2.5 million kilowatt-hours (kWh) of electricity in 1991. Just five years later, it had doubled production, won awards for quality, yet dropped its electric consumption to 1.7 million kWh, a remarkable one-third reduction. This was achieved with just prosaic improvements in energy efficiency such as variable speed motors that control the air in paint booths and changes in air compressor systems.[13] Toyota also moved in April 1998 to have almost all of its operations in Southern California, including its U.S. headquarters in Torrance, run by green energy, some 38 million kWh. As Romm puts it: "That Toyota is willing to pay a bit more for energy to reduce emissions is the strongest signal of the seriousness with which it takes global warming."[14]

This tend toward greener, carbon-reducing business has resulted from another, hopeful trend in the modern environmental movement—a greater focus on changing business through *both* punishment and reward. ExxonMobil is now the primary oil company that remains an obstacle to real change on global warming. Its obstinacy has led to campaigns against it that have their origins in the *Exxon Valdez* spill and in air pollution battles, but go much further. With positive federal change blocked, the Green Group organizations launched anticorporate campaigns against malefactors like ExxonMobil, while working with and encouraging greener or "cooler" companies like Toyota. In 2005, there was no controversy within the Green Group leadership about initiating its first-ever anticorporate campaign. It uses e-activism, media, and educational campaigns to call attention to the oil giant's desperate, last-ditch efforts to sabotage any progress toward reducing oil consumption and carbon emissions. Admittedly, the Green Group campaign, Exxpose Exxon,[15] is a milder version of Greenpeace's Exxon campaign, which also relies on direct action and demonstrations.[16] So, too, is an authoritative report by the Union of Concerned Scientists. It details and refutes Exxon's opposition to climate science.[17] Such targeting of corporate malefactors is a welcome innovation from established groups that had, until the Bush years, focused primarily on public policy.

Take for example, Defenders of Wildlife. Once a traditional conservationist group headquartered in a small brick townhouse with a wolf statue outside its Washington headquarters, Defenders evolved and grew rapidly under the guidance of its CEO, Rodger Schlickeisen. A direct-mail expert with a Harvard MBA and years of experience on Capitol Hill and the Interior Department, Schlickeisen is an avuncular, yet type A personality brought in during the early 1990s to revamp and expand the group. He did just that. Using advanced direct-mail techniques and modern marketing, Defenders grew rapidly. It now owns a big, glass-fronted green building in downtown Washington. I like to kid Rodger that his basic business is "selling wolves." But I mean it as a compliment. Defenders has not only helped to reintroduce the endangered gray wolf into the West over the stiff opposition of ranchers and political reactionaries, it has also been at the center of creating modern organizational mechanisms for the Green Group. These include mailing list enhancement, data mining, Web-based action campaigns, and more. Schlickeisen, with Gene Karpinski of US PIRG (now the president of the League of Conservation Voters), Bill Meadows of the Wilderness Society, and other forward- thinking leaders, helped round up widespread funder support and buy-in from all the major groups to carry out joint campaigns using pooled resources and agreed strategies. Called the Partnership Project, the new entity hired the Green Group's productive and popular coordinator, Julie Waterman, to lead it. The result has been effective targeting of legislative campaigns in local areas, joint efforts through SaveOurEnvironment.org, and more recently, the Exxon campaign.[18] In late 2006, for example, environmental groups in Washington discovered that former ExxonMobil CEO Lee Raymond had been selected by the Bush administration and its Department of Energy to head a task force to determine the U.S. energy future. The response was immediate. The Defenders of Wildlife, through its Web site and e-alerts, blared out: "These days, it's a lot easier to be an oil executive than a polar bear."[19] The group, whose sole mission had been defending such "charismatic megafauna," went on to warn that the task force would consist mainly of representatives of the National Petroleum Council. The action button was right beneath, of course, a picture of a polar bear. The significance should not be underestimated.

These environmental corporate campaigns, in turn, grow out of and are increasingly linked to long-standing activist concerns about corporate responsibility since the reform years of the late 1960s and 1970s. Consumer boycotts of Nestlé, GE, and other targets blended with shareholder actions, socially responsible investing led by nonprofit advocates, and centers for corporate responsibility. Signs of mutual interest were visible as early as 1991 when the Interfaith Center for Corporate Responsibility (ICCR) instituted shareholder actions calling for mandatory CO_2 reductions. The ICCR—an advocacy group consisting of more than 235 institutional, faith-based investors, including denominations, hospitals, and colleges with a combined $110 billion in investments—offers some two hundred shareholder resolutions per year. These call for responsible corporate behavior, including real action to stem global warming. Recently, the ICCR launched a joint project with Co-Op America calling for consumer action on climate change to affect the three large and popular mutual fund investors, Vanguard, Liberty, and American Funds.[20]

A New Local and Regional Focus

In addition to national efforts to prod corporate change on climate, environmentalists have moved to regional and local action and to cooperation and negotiation with friendlier corporations. A model for this sort of work has been Clean Air–Cool Planet, founded in 2000 by Adam Markham, a chipper British transplant to the States and longtime World Wildlife Fund (WWF) advocate at climate negotiations.[21]

Although Clean Air–Cool Planet is a regional New England NGO, its impact is broader; its annual meetings resemble national conferences. At one I attended in New York City, corporate, campus, environmental, and government representatives mingled happily. Although some corporate sponsors and participants are predictably small, green ones like Timberland outdoors products or Stonyfield Farms organic yogurt, there are far larger and paler green entities like Citigroup and Exelon, which owns and runs most of the nation's nuclear power plants. A combination of entrepreneurial spirit, marketing innovation, and passion about climate is a hallmark of the sponsors of Clean Air–Cool Planet. Stonyfield, for example, originally joined with the organization and with the Earth Day Network to form RenewUS.org, now called Climate Counts. It is a nationwide action campaign to get individuals to buy and use renewables. The campaign, launched in 2006, features a short, highly amusing Web-based video. Professionally produced, it shows what the future is like in 2055 when American citizens have demanded a brighter energy future. It looks amazingly real as you watch people pulling up to biodiesel stations, living in green buildings, and frolicking in clean air.[22] Like other organizations, CleanAir–Cool Planet also offers easy-to-buy carbon offset tags from Native Energy, a renewable wind power company owned and operated by Native Americans. These allow travelers to invest in clean energy projects to balance their carbon emissions. And these newer climate groups do not shun politics. CleanAir–Cool Planet testifies in Congress, issues the requisite policy reports, and attracts key politicians to its gatherings. At its annual conference, I was able to meet with Governor John Baldacci of Maine and political staff from most of the Northeast governors' offices. As I finished writing, Climate Counts (www.climatecounts.org) had expanded into a broad grassroots call to action. It offers sophisticated rankings and information about corporations for individuals to take action as both consumers and activists.[23]

Such efforts to go mainstream, rather than detracting from traditional grassroots efforts, have built on them, especially the long struggle for clean air. Impatient critics have accused the movement of going for small, incremental measures that, even if enacted, would not prevent the worst of climate change. To be sure, buying green power tags or Stonyfield yogurt, or waging a local battle against a coal-fired power plant will not, alone, prevent global warming. But both political change and grassroots organizing need clear focal points that affect people's personal lives. We need to feel that something small we do can actually make *some* difference. Few people can afford or tolerate the endless, tedious effort required to track and monitor the intricacies of global treaties, complex business deals, or scientific debates about climate modeling. But they will respond when their lives, their families, friends, neighbors, and other Americans are affected.

The struggle over coal-fired utilities is a good example. In Atlanta, local activist coalitions work closely with strong national organizations. They consist of religious groups, environmentalists, branches of public health groups like PSR and the American Lung Association, and more.[24] Their joint efforts have provided increased information, publicity and, ultimately, action against corporate giants like the Southern Company that have been harming humans and the environment. Their organizing has also been multiracial and has crossed class lines. Although filthy coal plants are commonly in poor communities and disproportionately affect minorities, the SOx, NOx, mercury, and CO_2 that they emit affect people regardless of boundaries—local, national, or global.

One of the most hopeful signs I witnessed at the Montreal climate negotiations following Hurricane Katrina was a delegation of local activists and officials from the Gulf Coast. They spoke to officials and the world press alike with passion and persuasiveness. Through a panel organized by the Southern Coalition for Clean Energy, the connections between CO_2 emissions, climate change, the growing destructiveness of hurricanes, and the impact of it all on human health and local communities was remarkably powerful.[25] The ability of the established environmental movement to grow and change could not have been clearer. Along with PSR's Ed Arnold and the chairman of Interface, Inc., Ray Anderson, one of the most effective figures was Jerome Ringo of New Orleans. Ringo is chairman of the National Wildlife Federation and president of the Apollo Alliance for Clean Energy and Good Jobs, a coalition of environmental, labor, and other groups with a bold policy plan for America's energy future.[26] A big, burly African American with a powerful voice to match, Ringo commands attention in any setting. He is about as far from Ranger Rick and the marketing of stuffed animals in NWF stores as you can get. I have been with Jerome in "insider" discussions with key senators, coalition meetings with Al Gore, and in international settings. His message is consistent. It is we humans, not just wildlife, who are harmed by pollution and climate change. It must stop. He tells audiences that minorities and poor people are especially hard hit. But, he says, there are other powerful reasons to shift to the bolder energy visions of the Apollo Alliance. These include more jobs, a stronger economy, and renewed moral leadership for America. For Ringo, there is no contradiction in the emerging new climate movement between impassioned grassroots and community organizing and close ties to the sophisticated policies, plans, and PR of established organizations.

FAITH-BASED MOVEMENTS

However we define the grassroots in America, they surely involve people of faith. The United States remains one of the most religious nations in the developed world. Churches are ten times more ubiquitous than Starbucks and McDonald's combined.[27] Other faiths like Islam or the Church of the Latter Day Saints (Mormons) are growing rapidly.[28] Historically, reform movements in this country—from abolition to women's rights, peace, labor, and civil liberties—have been spearheaded by Christians and Jews, as well as secular progressives. This is so true, in fact, that to this day, pronouncements or movements involving mainstream churches and synagogues are rarely treated as news.

I became aware of the reality of mainstream religious movements at a fairly early age, partly because I grew up in a family with many ministers in it. My mom went to New York Theological Seminary in the early 1960s. And my parents' minister from their days in Manhattan, Eugene Carson Blake, was prominent in Martin Luther King's 1963 March. He was a mainstay of the World Council of Churches, as was Cameron Hall, a friend and colleague of my mom's who attended our local church. My mother, Peg Musil, is descended from a long line of New England ministers, starting with Thomas Hooker, a founder of Connecticut, as well as reformers like the antislavery poet and activist John Greenleaf Whittier. During the 1950s and '60s, she became a kind of modern-day circuit rider throughout the Northeast. She trained teachers and Christian educators and worked with ministers from the suburbs to rural churches to the black congregations of Harlem and Bedford-Stuyvesant, bringing creative, innovative approaches to curriculum and the latest Christian concerns.

My former minister at the Garden City Community Church, Avery Post, is a learned liberal. He went on to head the national United Church of Christ. But not before he and other ministers in my youth group had introduced me to peace activists from New York, my first serious discussions with Africans struggling for independence, and a steady diet of theologically based concern for civil rights, tolerance, and justice. It was no surprise when I ended up a "youth representative" to the United Church of Christ's Task Force on Ministries to Military Personnel, trying to reform the Protestant chaplaincy and protest the Vietnam War.[29] Still in my twenties, I ran the Central Committee for Conscientious Objectors and was dealing with youthful draft and GI objectors and resisters nationwide. There I discovered, though we only discussed it privately, that my friend and colleague, Ron Young, head of Peace Education for the American Friends Service Committee (he had left his senior year at Wesleyan College to work for civil rights in Nashville alongside the renowned pacifist Rev. Jim Lawson) had in his youth, like me, attended the 1957 Billy Graham Christian Crusade in Madison Square Garden.

But by the '70s, young, liberal Christians like Young and me had become self-conscious about our faith and subsumed it into secular good works. Like Howard Dean or John Kerry in 2004, we did not wear our faith on our sleeves. Working amidst humanists, Jews, "red-diaper babies" who had grown up in the sectarian Left, and friends angry about their less liberal church upbringing, it would have seemed bad form to directly bring up Christianity. This hesitancy to mix admissions or pronouncements of faith with social action or politics carried well into the '90s, even as the so-called Christian Right gained media and political prominence. Nevertheless, I had been deeply moved, again and again, each time I attended some gathering where religion was mixed with social causes—from a 1971 antiwar GI service at the National Cathedral while I was an active duty captain in uniform, to masses based in liberation theology in Central America in the early '80s, to the sonorous sermons and snappy sound bites of the Rev. William Sloane Coffin, the civil rights and antiwar chaplain of Yale who went on to lead national efforts at New York's Riverside Church and the disarmament group SANE/FREEZE.

That is why I signed up for the first spiritual retreat held at the Fetzer Institute in Michigan that was offered to the Green Group by the Nathan Cummings Foundation. There I found myself looking at St. Francis of Assisi in a fresh, environmentalist light and meditating over passages from the Bible such as Jesus's somewhat cryptic admonition to "Consider the lilies of the field. Neither do they toil nor spin." Aside from the Sidney Poitier movie, I had mainly thought of this verse as a call to passivity or fatalism, not as a more spiritual and meditative call to perceive my own connection to the wholeness and goodness of God's creation. It was at this retreat, too, that I first met Paul Gorman, the head of the National Religious Partnership for the Environment (NRPE). Gorman had already been busy for years laying the foundations for renewed religious activism for the environment. We became close when he took the trouble to commute down from New York to weekly Sunday evening gatherings in Washington where a group of Green Group leaders met to try to pursue and deepen the spiritual foundations that had been stirred at the Fetzer Institute retreat. I found real community there, discovering that Phil Clapp, the sometimes scathingly sarcastic and brilliant head of the National Environmental Trust—the stereotype of a savvy Washington inside player—was a practicing Buddhist. Bill Meadows, president of the Wilderness Society, a wise and warm Tennessean with a leonine head, Roman nose, and curling silver locks about a high forehead that give him the appearance of a nineteenth-century senator, was a liberal Christian. Deb Callahan, the hyperenergetic, political, and eloquent leader at the time of the League of Conservation Voters, was, beneath it all, searching for deeper values and a spiritual center. But what I really found through the spiritual retreats and Sunday gatherings was my own voice. One of our group, Meg Maguire, then head of Scenic America and married to a Protestant minister, invited me to speak on the environment and public health at her church, the First United Church of Christ in Washington. What I recall at First Church was a packed room and an ease or spirit that came over me. I mixed imitations of Bill Coffin, biblical epistles, epidemiology, and something frighteningly like a modern-day altar call to give ourselves to saving God's creation. I was speaking with my whole being. I have thanked Meg many times for her faith in inviting me. And because she helped me knit together my spiritual and secular sides, the soulful and the scientific, into a public presence no longer afraid to speak about my faith.

A somewhat similar effect was produced on Carl Pope, the executive director of the Sierra Club who also attended the Fetzer Institute retreat and Green Group explorations of deeper values and vision. Though speaking from a moral perspective and not an explicitly religious one, Pope wrote an important, groundbreaking article in the *Sierra* magazine in 1998. He wrote about his own apology the previous year to the religious community for misunderstanding and stereotyping it. Pope had spoken at a symposium in Santa Barbara hosted by the patriarch of the Orthodox Church, Bartholomew I. The head of the Sierra Club did not suggest that environmentalists needed necessarily to be religious, but that they had "ignored to our detriment the power that organized religion can bring to our mission." He mentioned specifically the "tremendous conviction and enthusiasm with which the National Council of Churches has approached global warming as a moral issue, or the fervor and effectiveness of the Evangelical Environmental Network in speaking out for the Endangered Species Act."[30] But what was extremely brave and unusual for an environmental leader was Pope's analysis of the effect on him and a generation of activists of a popular late-1960s essay by UCLA historian Lynn White Jr., who argued that the Judeo-Christian tradition had driven Western civilization to devalue the natural world. He had asserted that "we shall continue to have a worsening ecologic crisis until we reject the Christian axiom that nature has no reason for existence save to serve man."[31] Upon rereading White's essay, Pope realized that he and other environmentalists had misread the historian. He actually noted that Christianity is far from monolithic and has many gentler, ecological strains. "We became," Pope writes, "as narrow-minded as any religious zealot."[32] Today, miraculously enough, the Sierra Club actually has a staff position and a "Faith and the Environment" partnership devoted to working with the religious community. On its Web page, the Club notes that "67% of Americans say they care about the environment because nature is God's Creation." Some 47 percent of Sierra Club members say they attend religious services at least once a month and the Club "empowers [them] to partner with their local faith communities to educate the public about solutions to America's environmental challenges."[33]

It should be reason for hope, too, that efforts by mainstream religious groups, long ignored or taken for granted, are now bearing fruit. Paul Gorman described the recent outpourings of both evangelical and liberal Christian caring about global climate change as "a culminating moment of awakening of the real concern for environmental issues that began as early as 1990 across denominations and varieties of religious life."[34] The sheer magnitude of climate change affects all of God's creation on earth, according to Gorman. It integrates long-standing religious concerns for children, the poor, and social justice. The climate movement, he says, has become "a vessel for all that is felt for the urgency, the moral and political power, of what people of faith feel for the depth and scale of the cosmos."[35] Gorman, winner of the Heinz Foundation Environment Prize, and a civil rights veteran, speaks by turns in a soft, quizzical, Socratic manner, followed by telling, Talmudic explanations, mixed with TV sound bites. The new religious climate movement, he tells me, with a twinkle in his voice, allows religious people to talk about and extend their moral concerns, "to the cosmos and not just the crotch!"[36] Gorman worked with various leaders of religious denominations to found the National Religious Partnership for the Environment between 1991 and 1993. Drawing on expressions of social concern going as far back as the postwar founding of the World Council of Churches and various social concern ministries of the National Council of Churches, Gorman spent much of the time discussing and deepening the relevance of such broad outpourings. The founders of NRPE drew on the views of the World Council in its statement, "Justice, Peace, and the Integrity of Creation," and similar pronouncements. As Gorman puts it, he was working with "the prophetic minority who saw in the biblical case for the poor a powerful opportunity to take action and build a new movement that would bring a powerful new constituency to the environmental movement."[37] He recalls, among other insights, a moment when Roman Catholic Bishop James Malone of Ohio, who had been involved with economic justice and workers rights in the dying plants of industrial Youngstown said to him: "Paul, why don't environmentalists ever have pictures of *people?*"[38]

Linking the environment, science, and concern for people led Gorman in 1991 to initiate a "Summit on Religion and the Environment." It followed an open letter signed by the Cornell University scientist Carl Sagan, of *Cosmos* fame, and numerous luminaries from science and religion.[39] The summit put religion on the environmental map and in the public eye for the first time. And it led, ultimately, to the formation in 1998 of the National Council of Churches' Interfaith Climate Campaign. Throughout, Gorman and others cultivated evangelical Christians, believing that a careful examination of their faith and scripture would lead them to care about creation. They would become a counterweight to the politicized right-wing clerics and their allies who saw environmentalism as Satanism. The Interfaith Climate Campaign worked in eighteen states, relying on networks from the National Council of Churches, mainline denominations, and local churches and synagogues to educate, sermonize, and organize in the grass roots. Like secular movements, the religious campaign often organized on issues that touched people's immediate lives or were relevant to legislative battles at hand for clean air or CAFE standards.[40] Pointing out that SUVs were exempted from national gas mileage standards, the group produced a TV ad and sound bite that in many variants, has entered our lexicon. "What kind of car would Jesus drive?"[41] It resonated from the lips of politicians, commentators, comedians, and people across America. Environmentalists had, at last, broken the taboo that had kept them apart from millions upon millions of ordinary Americans. It was the mention of Jesus, the humor, and the involvement of evangelicals alongside mainline religious liberals that got the media's attention. So had the formation of the Evangelical Environmental Network (EEN). EEN puts out *Creation Care* magazine and has the support of numerous heads of Christian colleges, seminaries, and Bible institutes.[42] Their biblical foundation is the familiar, direct, and sometimes neglected verse, Psalms 24:1: "The earth is the Lord's and the fullness thereof." What is critical about this constituency, aside from its surprise or news value, is its potential political importance. Whereas African American evangelicals tend to vote Democratic, white evangelical Christians make up about 23 percent of both the U.S. population and the electorate. In 1987, they were split about evenly between Republicans (34 percent) and Democrats (29 percent). Today, 48 percent of white evangelicals identify as Republicans. Of those who voted in 2004, 78 percent went for President Bush, making up 36 percent of his total.[43] But growing awareness of hunger, poverty, overseas genocide, and now the reality and threat of global climate change and oil dependence may begin to destroy that political nexus. Signs of how this might occur include the growing size and sophistication of operations like the Sojourners Community led by the Rev. Jim Wallis. Beginning as a Christian evangelical magazine concerned with social justice and peace, Sojourners now boasts an e-activist network of 2 million members and mounts regular lobby days. It spreads the word through books and media appearances by Wallis, whose resonant bass voice resembles that of the stereotypical patriarchal God.[44] In addition to EEN and Sojourners, another group of evangelicals concerned with climate made news in 2006 when eighty-six influential leaders of the evangelical community signed a widely publicized statement calling for "federal legislation that would require reductions in carbon emissions through cost-effective, market-based solutions." Signers included new voices like Rick Warren, author of the hugely influential *The Purpose Driven Life* and other books; Jesse Miranda, president of AMEN; and influential black evangelical leaders like Bishop Charles E. Blake Sr. Formed into the Evangelical Climate Initiative, the group has aired politically oriented television spots in Republican or closely competitive states such as Arkansas, Florida, Kansas, New Mexico, North Carolina, South Carolina, South Dakota, Tennessee, and Virginia.[45]

The real news here is that mainstream religious denominations and leaders are now working alongside evangelical ones. As a result, they are helping to get the attention of the media and politicians. Paul Gorman, working with evangelical leaders, was instrumental in the spring of 2005 in instigating an influential two-and-a half-day Capitol Hill conference plus lobbying sessions held by the National Association of Evangelicals. It focused on climate change. I was surprised to learn that the group was addressed by Sir John Houghton, the chairman of the Scientific Assessment Panel of the Intergovernmental Panel on Climate Change (IPCC). Houghton had personally briefed Margaret Thatcher, guided the deliberations of the world's scientists that undergird the Kyoto Protocol, and is the author of the single best science review of climate, *Global Warming: The Complete Briefing*. Only this time, Houghton spoke as a committed Christian. After laying out a concise summary of climate science, Houghton told his audience: "a particularly strong challenge to our stewardship comes from the realisation that the adverse impacts will fall disproportionately on the poorer nations. . . . We in the developed countries have already benefited over many generations from abundant fossil fuel energy. . . . There is already a tendency for the rich to get even richer while the poor get poorer. The impacts of human induced climate change will tend to further that trend. Let me remind you of the words of Jesus spoken after he had told the parable contrasting the faithful and the unfaithful stewards, 'For unto whomsoever much is given, of him shall be much required.'"[46]

How important are such linkages? It is hard to measure, but Republican Senator Lindsay Graham of South Carolina, a Baptist, thanked the Christian climate lobbyists and told them that such visible manifestations "help give him the political cover he needs to increasingly speak out and support efforts to halt climate change."[47] More recently, Mike Neuroth of the United Church of Christ organized the National Council of Churches Ecumenical Advocacy Days for 2007.[48] Some one thousand Christians traveled to the Capitol to learn about and lobby for peace, justice, and the prevention of climate change. This was followed by Earth Day 2007. The secular Earth Day Network provided Christian materials, Web sites, organizing, and inspiration that led on Sunday, April 22, to some ten thousand to twelve thousand Earth Day sermons in churches across America. That's as if we *doubled* the number of seemingly ubiquitous Starbucks in our land and held a religious Earth Day service in every single one! And, of course, there were secular Earth Day events and rallies around the nation as another one thousand lobbyists descended upon Washington.[49] Other connections between secular and spiritual environmental movements are growing, too. In early 2007, the Sierra Club and the Massachusetts Council of Churches together organized the first statewide march of environmentalists and religious bodies from western Massachusetts to the capital in Boston. Their event mimicked Bill McKibben's precedent-setting grassroots Vermont climate change march, which went on to give birth to the "Step It Up" campaign.[50] In April 2007, "Step It Up" ultimately turned out more than 1,400 local climate change events at environmentally significant places, along with a culminating human montage in Washington. At each "Step It Up" event activists were carefully instructed also to contact their representatives and to take *political* action as well. Given the campaign's success, McKibben planned "Step It Up2" scheduled one year before the election of 2008. It, too, proved a nationwide success and joined in the nation's capital with PowerShift 2007, a youth summit on climate with more than 5,500 attendees, 1,000 of whom stayed on to lobby.[51] The line between faith-based efforts and traditional grassroots activism is becoming blurred in positive ways, especially when we note that McKibben is also a lay leader in his local Unitarian Universalist church, writes sermons, and can be found in the pages of *Christian Century,* as well as in the online environmental magazine *Grist!*[52]

Religious environmentalists have also looked to the mote in their own eye, launching organizations to reduce the carbon emissions and toxic wastes emanating from all those churches and synagogues (and some mosques and ashrams as well) that dot the American landscape. Interfaith Power and Light (changed to InterfaithWorks in 2006), for example, proselytizes nationwide (OK, they educate and train) about the virtues of switching to green, renewable power sources and using efficient and low-carbon-emitting lights, heat, and air-conditioning. They also use their growing size and connections to help churches and individual parishioners link to green power, to incorporate sustainability into their own homes and lives, and to spread the word.[53] Shortly before the 2006 election, for example, Interfaith Power and Light provided educational kits, talking points, and a DVD of *An Inconvenient Truth* to more than four thousand gatherings throughout all fifty states.[54]

CAMPUS AND YOUTH CLIMATE MOVEMENTS

If you want to have hope for the future, just spend some time with today's youthful leaders of the peace and environmental movements. They are idealistic, politically savvy, energetic, and even more effective than those who protested the Vietnam War during the 1960s. The organized networks of student branches of the main Green Group organizations alone communicate with and organize nearly a million students. And that is just the tip of the iceberg. Like much of the environmental movement, campus movements, which are decidedly grassroots, are fairly decentralized. They focused initially on more traditional issues like air quality, toxins, and natural resources. But as awareness of globalization and inequalities have grown, students have mounted anticorporate campaigns aimed at multinational polluters, in support of global workers, and, in recent years, for campus sustainability and global climate change.

Free the Planet is at the core of the new youth environmental movement.[55] Founded in 1995 in response to the Gingrich "Contract with America" and its antienvironmental sweep of the 104th Congress, students decided to fight back. More than one thousand students from fifty states issued "A Call to an Emergency Environmental Conference." In February, 1,800 students gathered in Philadelphia for the first such student-initiated national action. It was sponsored by existing student environmental groups like the Sierra Student Coalition, the Student PIRGs, Campus Green Vote, Green Corps, the Student Environmental Action Coalition (SEAC), and others. Soon 450 campuses were involved. They produced three hundred thousand student signatures, including that of Newt Gingrich's niece. Students demonstrated, staffed sixty-six canvass offices, and, between spring and fall 1995, collected more than 1.2 million signatures for an Environmental Bill of Rights Petition, with five hundred organizations joining the campaign. The petitions were delivered to Capitol Hill, and the next day Congress voted to strip seventeen antienvironmental riders from the EPA Appropriations Bill. As Free the Planet, students collaborated with national

Green Group organizations and members of Congress, even as they carried out dramatic actions like spearing a scale model of the earth that had been produced for and was to be prominently displayed on Earth Day by corporate polluters seeking to improve their image. The protest gained wide publicity; the polluter-sponsored event was cancelled.

Free the Planet has gone on to train hundreds of young environmental activist leaders and carry out campaigns such as one aimed at Office Depot, the world's largest office supply store. It led to the company's using only recycled paper instead of paper coming from rare and vulnerable forests, monoculture tree forests, or genetically modified trees. The same is true for a Citigroup campaign that included hundreds of campus events, some confrontations with the bank, and twelve thousand individual pledges never to do business with it unless they disinvested from certain kinds of mining, logging, or intrusive operations in primary tropical forests. On January 22, 2004, as a result of the campaign, Citigroup announced a new environmental investment policy. It became the first multinational bank to prohibit investment in any extractive industry. Similar campaigns were carried out with targets like Boise Cascade, Staples, Pepsi, and Home Depot.

Beginning in 2001, through Free the Planet, local campuses took actions to ratify Kyoto and to switch campuses to wind and other renewable power. By 2006, such efforts had led steadily to more and more interest and action on global climate change. First came Sustainable Campuses, a campaign to decrease the environmental impact of colleges and universities. Next was Cool the Planet, a campaign to stop global warming. It has included initiatives like Drive Ford Green, designed to push the Ford Motor Company to move toward lower emissions, and political campaigns aimed at pressuring the Bush administration to cut climate change emissions.[56]

Increasingly, large and focused efforts on climate change and energy have emerged on campuses. A particularly energetic leader is Billy Parish, who left Yale to coordinate Energy Action. It is a coalition of more than thirty campus and youth-oriented groups in partnership with established campaigns like Free the Planet and the Climate Campaign. It reaches 539 campuses and includes activist voter education groups formed during the 2004 campaigns like Americans for Informed Democracy, and Campus Progress, the youth wing of the new activist think tank, the Center for American Progress (CAP).[57] This growing environmental youth and campus initiative also includes more diverse groups like the Indigenous Environmental Network and the Environmental Justice and Climate Change Initiative, along with professional types like the National Association of Environmental Law Societies and Student Physicians for Social Responsibility. Again, at the core of such new efforts are the student and youth wings of established groups like the Sierra Club, the PIRGs, Greenpeace, the League of Conservation Voters, the National Wildlife Federation, and others.

The Energy Action coalition reflects the more sophisticated style of principled yet pragmatic youth in contemporary movements. It implicitly recognizes that their elders in the '60s and '70s too often avoided engagement in the political arena. They shunned techniques and styles that seemed too "bourgeois." In its vision, Energy Action focuses on four strategic areas: "Campuses, communities, corporate practices, and politics." In 2006, it formed and issued the Campus Climate Challenge, whose goal "is to build a generation-wide movement to stop global warming. . . . By empowering the generation most threatened by global warming to be front and center in *solving it* [emphasis added], we can make this an urgent and mainstream issue for society as a whole. The Campus Climate Challenge, designed to run until the inauguration of a new President in 2009, talks about young people as 'the most sought after slice of consumers in America. Their tastes and interests dictate ad campaigns, story lines on sitcoms.'"[58]

Student groups and leaders on climate change and other energy issues also work with faculty and administrators. At Lewis and Clark College, for example, economics professor Eban Goodstein formed Focus the Nation. It is an initiative specifically aimed at putting the educational and political spotlight on climate change through events that led up to a nationwide teach-in on over one thousand campuses on January, 31, 2008.[59] When I spoke at St. Mary's College in California in February 2008, the campus there was still abuzz over the Focus the Nation event whose primary sponsor and contact was Dean of Liberal Arts Stephen Woolpert. St. Mary's had held a day-long Eco-Fair with green vendors, advocacy groups, and films. An evening symposium featured national experts like Stephen Schneider, codirector of the Interdisciplinary Program in Environment and Resources at Stanford and Dan Lashof of NRDC. But newer educational efforts like Focus the Nation also draw on action campaigns and marches. In 2004, after years of campus sustainability organizing, the twenty-three college chapters of MASSPIRG ran a statewide "Stopping Global Warming Starts Here" campaign involving more than ten thousand students. It convinced then Governor Mitt Romney to adopt a fairly strong climate action plan.[60]

In fact, such organizing drew me in October 2006 to a Focus the Nation conference at Middlebury College where Bill McKibben has been a scholar in residence. Middlebury, and its leading environmentalist faculty member, Jonathan Isham, a professor of international environmental economics and former World Bank economist, had already hosted a 2005 conference calling for a new climate movement. That gathering and *Ignition,* the book written after it, has helped further stimulate a growing nationwide trend toward collaboration between college administrators, faculty, and students.[61] Now, almost half of the colleges and universities in the country have initiated sustainability initiatives of some kind, one hundred of which are buying or generating more than 500,000 mega-watt hours of clean energy annually. It is the largest single driver of the voluntary clean energy market.[62]

The National Wildlife Federation's Campus Ecology Program has been a pioneer in this movement. It has developed a base of engaged faculty and facilities managers who have been transforming campuses.[63] These efforts, growing out of local initiatives and national leadership have continued to spread. They have helped spawn entirely new, specialized campus groups such as AASHE, the Association for the Advancement of Sustainability in Higher Education, the National Council on Science and the Environment, and others. By 2007, AASHE had already signed up more than three hundred college presidents to a pledge to make their campuses sustainable and carbon neutral. Even a well-established national higher education association like the 1,100-campus member Association of American Colleges and Universities (AAC&U) now has regular sessions on sustainability at its national conferences.[64]

Cool Cities

The new climate movement has also sought influence in cities and towns. The result has been city campaigns linking activists with mayors and city councils. Seattle Mayor Greg Nickels initiated the U.S. Mayors Climate Protection Agreement on February 16, 2005, the same day that the Kyoto Protocol took effect in 141 nations. By the 2006 elections, more than 330 mayors, representing 53.3 million Americans in some forty states, had signed on. The pledge commits their cities to meeting the initial Kyoto goals for the United States—a 7 percent cut below 1990 levels by the year 2012—even if the federal government refuses to participate in that agreement. Nickels now speaks and travels far beyond Seattle. He played a leading role at the Montreal climate negotiations in late 2005 and at a major conference there sponsored by the International Council for Local Environmental Initiatives (ICLEI), which formed a World Council of Mayors.[65] At home, Seattle has already reduced global warming emissions by more than 60 percent through green vehicles and buildings and hosted the November 2007 U.S. Conference of Mayors Climate Protection Conference, attended by Al Gore, Bill Clinton, and more than one hundred mayors. Seattle City Light led the way as the first electric utility in the nation committed to net zero carbon emissions. The city has established a Green Ribbon Commission on Climate Protection where leaders from the environmental movement sit side by side with government, business, labor, and community representatives. Nickels has also initiated an ambitious tree-planting program. In September 2006, he announced a plan to maintain Seattle's current tree cover while planting an additional 649,000 shade trees over the next three decades. Like many cities, Seattle had lost more than half of its tree canopy by the early '70s as population and development spread.[66]

Similar efforts are now under way in cities that have lost shade cover such as Washington, Baltimore, Minneapolis, Chicago, Denver, and Los Angeles. During the torrid summer of 2006, Sacramento set records for heat, with eleven straight days over 100°F. The city plans to plant at least four million trees in addition to the 375,000 shade trees they have given away to residents over the past sixteen years. Residents of the California capital simply call up the Sacramento Municipal Utility District, a publicly owned power company, to receive up to ten free trees.[67] According to the Department of Energy, shade trees planted strategically around a house can reduce air-conditioning bills by about 30 percent in hot, dry cities like Sacramento. In Los Angeles, where average evening temperatures have risen by 7°F over the past century, a campaign has started to plant a million trees, based on the Sacramento model. According to Paula Daniels, a commissioner on the Board of Public Works, Los Angeles can expect to get a $2.80 return on every dollar it spends on trees from energy savings, pollution reduction, storm-water management, and increased property values. It's easy to see why. A Lawrence Livermore National Laboratory study that shows that L.A. could, by planting 10 million trees and making lighter-colored roofs and pavement, reduce the "heat-island effect that has been growing the past hundred years and lower its peak summer heat by five degrees, cut air-conditioning by 18 percent and reduce smog by 12 percent."[68] And, of course, CO_2 emissions would be reduced accordingly. The group American Forests has been running extensive campaigns to get trees back into cities and offers a link on its Web site on how to get them and what the effect on carbon emissions will be. That's partly because federal funding for the Urban and Community Forestry Program was cut by 25 percent during the first four years of the Bush administration. Small to begin with, total federal spending for trees that improve neighborhoods, save energy, and cut climate emissions sank from $36 million annually to a proposed $27 million for fiscal year 2007.[69]

But city efforts are not limited to tree plantings, nor to cities in blue states. Bill White is the mayor of Houston, Texas, one of the nation's most polluted cities. In the spring of 2005, he announced that the entire city fleet of cars, pickup trucks, and SUVs would be converted to hybrids. Eighty percent of all new purchases and 50 percent of the entire fleet will be using hybrid engines by 2010. In the nation's fourth-largest city, that's a powerful political message and a huge dent in carbon emissions. A single Toyota Prius, for example, will, over a lifetime, emit 43 tons fewer greenhouse gas emissions than a conventional car.[70] Incredibly enough, one of the most popular and successful mayors who proselytizes about global warming is Rocky Anderson. He is the Democratic mayor of Salt Lake City, the capital of one of the most conservative and "red" states in the nation. Anderson has turned climate change efforts into a political gold mine. Salt Lake City has already exceeded Kyoto targets for emissions cuts, and as Anderson says: "When I can get up in front of the Salt Lake City Rotary Club, which is by and large conservative business people, and get a standing ovation after talking about the kinds of changes we're making here, that says a lot."[71] All this local action has been stimulated and backed up by environmental activists working outside the limelight. Take a program like the Sierra Club's campaign, "Cool Cities: Solving Global Warming One City at a Time." Citizens can get a free guide along with news and updates on city efforts to fight global warming. With more than 750,000 nationwide members, local chapters, and national analysts and organizers dedicated to the campaign, the

Sierra Club maintains sustained pressure through coordinated strategies at the local, national, and international levels.[72]

STATE ACTION

States are also centers of new action with heavily populated East and West Coast areas leading the way. An initial stimulus came from then Governor George Pataki (R-NY), when he initiated the Regional Greenhouse Gas Initiative (RGGI) process in 2003, following George Bush's withdrawal from the Kyoto Protocol. By 2006, the RGGI states included New York, Connecticut, Delaware, Maine, New Hampshire, New Jersey, and Vermont. In summer 2006, the RGGI states announced, following lengthy research, debate, and their own negotiations over final levels, a rule that would create a market for CO_2 emissions from the Northeast region's power plants. Although the initial rules proposed were mild, capping CO_2 at current levels (no growth) from 2009 to 2015 and cutting them 10 percent by 2019, the agreement reached by a bipartisan set of governors sent a strong political signal. The caps were tough enough, however, to have led to the withdrawal from discussions by Massachusetts and Rhode Island.[73] They did not join RGGI until 2007, following significant pressure from environmentalists. Then Maryland followed suit after adopting clean air legislation capping carbon emissions. That bill also was the result of an intense campaign by environmental groups who got sufficient backing to prevent outgoing Republican Governor Bob Ehrlich from vetoing it.[74]

Southwest and Northwest governors, including New Mexico's Bill Richardson, have joined in the climate action, too. But the megastar of state climate efforts has been Governor Arnold Schwarzenegger (R-CA). His climate change reforms and opposition to Bush administration policies saved his flagging 2006 gubernatorial reelection campaign. In August 2006, the California legislature, after last-minute bargaining, signed a law that made California the first U.S. state to mandate cuts in greenhouse emissions equal to 25 percent by 2020. Given the size of the California economy and Schwarzenegger's high profile, the initiative drew worldwide attention. It raised the standard yet again for national politicians of both parties.[75] But like most such news-making breakthroughs, work by national environmental organizations linked to local, grassroots efforts lay beneath the headlines. The Schwarzenegger cuts and the legislation mandating them were introduced by Assemblywoman Fran Pavley along with speaker Fabian Núñez. Pavley worked closely with both Environmental Defense and the Natural Resources Defense Council, which cosponsored and worked on the bill, just as they had with pioneering auto emissions standards sponsored by Pavley in 2002.[76]

National and Grassroots Groups Build New, Unified Campaigns

It was, in fact, Environmental Defense, NRDC, and the National Wildlife Federation who in 2005 took the lead in developing a truly nationwide campaign associated with former Vice President Al Gore. Each of these groups and the entire Green Group of environmental organizations had decided as early as 2002 that global climate change needed to be their top priority. Following overtures and work with labor, religious groups, environmental justice organizations, public health, and state and local coalitions, the Green Group examined its own successes and failures, along with research and advice on framing and communications. Then, following the narrow loss by John Kerry to George Bush, the Green Group agreed unanimously to build an even larger climate campaign, whether or not outside funding could be found. The goal was to make climate a major issue for the American public by the time of the 2008 elections. Meanwhile, a separate initiative being designed and pushed by Michael Northrop of the Rockefeller Brothers Fund was taking shape. Its initiators were talking to various large funders. They eventually approached Al Gore. Gore was already carrying out his now famous lectures and had access to additional large donors. After a lifetime of public service, he had also managed to earn significant money and was eager to help. But he made it clear that any campaign should not be seen as solely identified with him. Importantly, in private conversations, he added that he would not participate if the Green Group organizations were not included.

The new coalition opened its doors in 2006 as the Alliance for Climate Protection.[77] Gore provided serious seed money by donating royalties to the campaign from his best-selling book and documentary, *An Inconvenient Truth*. Initial planners and participants ranged from JP Morgan Chase Bank to the Urban League, the National Catholic Rural Life Conference, the Garden Club of America, the United Auto Workers, the Evangelical Environmental Network, as well as traditional environmental groups, science and public health groups like the Union of Concerned Scientists and Physicians for Social Responsibility, and campus groups like Energy Action. Determined to be inclusive, broad-ranging, and bipartisan, the group formed a board of directors that included prominent Republicans such as former National Security Advisor Brent Scowcroft. Meanwhile, Green Group members and other organizations worked with Paramount Pictures to draw people to *An Inconvenient Truth* and to build events around it. Thus, it was not exactly spontaneous when the Gore film shot out of the box office starting gate, moved to a larger number of theaters, and leapt to near record attendance.[78] Gore had been convinced to do the film by Laurie David, an environmental activist and then wife of television star and producer Larry David. She had already created a novel organizing approach to climate change called the "Stop Global Warming Virtual March." The virtual march had signed up more than half a million e-activists by 2006 through viral marketing, the involvement of recognizable experts and celebrities, and other modern techniques of Web-based activism.[79] With David Guggenheim as producer, the film engaged other celebrities and surprised audiences with its stunning visual science lectures, Gore's "new" personality, and its timely, compelling information. Gore went on to gain an Oscar, and the coalition trained and supported more than one thousand speakers to present the global warming slide show. One of them, Gary Dunham, is a seventy-one-year-old Republican whose wife belongs to the Daughters of the American Revolution. He came upon a discussion of the Gore film while in a motel, channel surfing past the Oprah Winfrey show. He took out the film, watched it twice, and soon headed off to be trained. He is now part of the Gore operation and has given the slide show to U.S. troops in Iraq. In 2007, the former vice president and the coalition then mounted simultaneous "Live Earth" global warming concerts and events around the globe.[80] Finally, in Oslo in October 2007, Gore, along with all of the 2,400 scientists of the IPCC, had their tireless work, so long ignored or considered controversial, recognized by the Nobel Prize for Peace. By the end of 2007, Gore's group was joined by the launching of another new national effort, the 1Sky Campaign. Growing out of the initial explorations of the Rockefeller Brothers Foundation and its network of concerned organizations and activists, the campaign was designed by a team of activists and funders headed by Betsy Taylor, a former foundation official and head of the Center for the

New American Dream. 1Sky incorporated activist entities like McKibben's "Step It Up" and PowerShift2007 and locally based efforts like the Southern Alliance for Clean Energy and the Chesapeake Climate Action Network. The 1SkyCampaign offers a bold vision and program. But unlike many previous grassroots initiatives, 1Sky has been careful to include and coordinate with the grassroots efforts of traditional environmental groups and announced its determination to affect federal policy and to raise climate change issues among candidates.[81]

Growing Political Action

The emergence of bold, grassroots, yet politically engaged and savvy groups like 1Sky willing to engage with Washington politics and the established Beltway organizations may signal the crumbling of the final obstacle to a truly strong environmental and climate change movement. It has been a long struggle to get enough national and local environmental activists and organizations to put aside internecine warfare, work together, and move beyond education and organizing into sophisticated lobbying and electoral politics. Until the mid-1990s, criticisms of the mainstream national movement as disengaged from politics or following mainly elite approaches had some merit to them. But the 1994 Gingrich revolution was a wake-up call to build unity, to work with and develop more grassroots groups, and to get serious about fighting back, especially in the electoral arena. With increasing momentum, traditional Green Group environmental organizations began changing fairly quickly. The narrow Gore defeat—after a monumental and largely successful effort to identify, engage, and turn out environmental voters—led to record environmental electoral efforts in the 2004 election. John Kerry, backed by environmental organizations, fell just short of victory with more votes than any Democrat in history. He also drew record youth votes—against a then popular incumbent wartime president.[82] But once again, many funders and some in the grassroots movement were despairing or angry. But by 2006, most had seen the horrendous effects of a totally antienvironmental government. They opted to give even greater priority to influence through electoral politics.

The result helped sweep out the antienvironmental Republican Congress. Deb Callahan, who built the League of Conservation Voters (LCV) into a powerful electoral machine that educated, trained, and worked with local activists nationwide, had for years called for a more "political" environmental movement. After a second narrow defeat with Kerry, Callahan, like other leaders, took a break from the movement after ten years of virtually round the clock effort. But the LCV, the electoral arm—along with the Sierra Club—of the Green Group organizations, built on the infrastructure and momentum she created to achieve an incredible winning record in 2006.[83] Far beneath the notice of major media or even savvy political pundits, LCV, the Sierra Club, and other groups collaborated on educational work, voter registration, turn-out operations, advertising, and plain old retail, grassroots politics to swing critical electoral districts toward environmental candidates. The lines between "hard-core grassroots activist" and Beltway-bound bandits were truly blurred. The day had arrived when the Sierra Club, the Garden Club of America, Greenpeace, graying Clinton officials, and local, grouchy greens could—like the lion and the lamb—lie down together. By 2008, the broader environmental movement was deeply involved in electoral races, presidential primaries, and pushing climate change.

In 2008 and long after, global climate change must become and remain a key electoral issue for our representatives—from city councils and mayors, to governors, members of Congress, senators, on up to the presidency. There is an opportunity for every citizen to make sure that our elected leaders put a new, healthy energy future and preventing global climate change at the top of their agenda. The economy, health care, education, and, of course, national security remain vital. But the time has come, after a truly dark period, for a new day. The time is short. As scientists and environmentalists like to point out, the planet does not care whether we take action. Like Noah and his wife and others who have followed, it is all up to us.

What is needed is an end to false distinctions between the personal and the political, the grassroots and the Green Group, the spiritual and the secular. We need credible and clarion calls to stop global climate change, create new energy policies, and begin to live our personal and political lives in new ways. In my final chapters, I will lay out a simple set of policy prescriptions to take to the grass roots and to candidates, as well as a set of actions that you, your family, and friends can take at home and in your community. They will begin to reduce carbon emissions, connect you to like-minded people and organizations, influence our leaders, and give you, and the rest of us, real hope.

7

Where Do Emissions and Energy Come From?

In this chapter, I want to look at where and how we get our energy, trends in its use, and the prevailing, conventional assumptions about why our future will continue to be dominated by fossil fuels. But our future is not inevitable. It depends on changes in the political climate and on our own actions. When we fast forward to 2030 or 2050, no one can be certain what our energy needs and sources will be. But I do know that what we've got is inefficient, costly, and dangerously unhealthy. Our energy sources are outmoded; they threaten to destroy our way of life. But given the economic and political momentum behind our current fossil fuel and nuclear economy, we need to understand its deficiencies if we hope to convince others to switch.

How We Use Energy and Where It Comes From

In the introduction, I talked about how much I love books. That's why I empathized with Gus Speth, dean of the School of Forestry and Environmental Studies at Yale, when I searched for his book, *Red Sky at Morning.* It ought to have been a best seller, not just an environmental policy classic. I found it in my local bookstore, as he had ruefully predicted, jammed in a corner with the cat and dog books, bird guides, and nature essays. I could claim to have strolled into Barnes & Noble to get Speth's book. But the truth is I drove the mile and a half to downtown Bethesda. I circled the blocks searching for parking, entered a huge, brightly lit, air-conditioned store, bought a Starbucks coffee with beans picked, shipped, roasted, and packaged God knows where, returned to my air-conditioned study, flipped the switch to power up the lights, the computer and printer, and promptly headed to the fridge for my first break. All this took energy. Lots of it. Not mine. I didn't break a sweat. But the stuff of our civilization: the paper, printing, advertising, shipping for those books, the air-conditioning, the cars, the food flown in from afar and kept cold in the fridge, the furniture, lights, the Internet and phone, the nicely paved roads to and from my brick colonial home—all take energy to produce and use. And right now, in the United States, most of that energy comes from fossil fuels—oil, gas, and coal. Some is from nuclear power. But only a tiny bit comes from renewable sources like wind and solar energy.

You and I and the rest of the citizens, or consumers, in the United States rely on fossil fuels for 86 percent of our energy. The Energy Information Administration (EIA) tells us that 40 percent overall comes from oil, 23 percent from natural gas, 23 percent from coal, 8 percent from nuclear power and the remaining 6 percent from hydro power and renewables.[1] What's confusing is that we use different kinds of energy for different things. Most of the oil we use, for example, goes toward transportation. It powers cars, trucks, and planes. In the United States the bulk goes to autos. So, if we want to reduce our dependence on foreign oil, it's simple. We've got to create more efficient cars, use other fuels, switch to electric cars, take public transit, and drive less, or some combination. No amount of changing your light bulbs or saving on electricity will help.

Electricity comes from coal, natural gas, and nuclear-powered generators (and that little bit from wind and solar power and other renewable sources.) If you want to get rid of coal-fired utilities or nuclear power plants, you need to switch to electricity produced by renewables, or use less in your home, office building, or place of worship. No amount of bicycling or taking the bus to work or car pooling will help. (About 70 percent of our electricity comes from fossil fuels, mostly coal and gas, 21 percent from nuclear plants, and only about 9 percent from renewables.)[2]

Despite some improvements and efficiencies, energy usage in the United States continues to creep upward at about 1 percent annually as our population, economy, and consumption have increased. Each year, we release about 6 million metric tons of CO_2 emissions into the atmosphere.[3] Let's take a look at the fuel sources that give off carbon and difficulties with them. There are limitations, environmental and health impacts, and carbon emissions from all fossil fuels. That's why some analysts suggest we need to restart our nuclear power industry. I disagree. I have included nuclear power here as one of the energy sources we need to avoid—along with oil, gas, and coal.

Oil

First let's look at oil. As Jeremy Leggett, now a solar energy entrepreneur, puts it: "We have allowed oil to become vital to virtually everything we do. Ninety percent of all our transportation, whether by land, air or sea, is fueled by oil. Ninety-five percent of all goods in shops involve the use of oil. Ninety-five percent of all our food products require oil use. Just to farm a single cow and deliver it to market requires six barrels of oil, enough to drive a car from New York to Los Angeles."[4] That means that the world consumes more than 80 million barrels of oil per day, 29 billion barrels per year. The International Energy Agency, set up by the major industrialized nations, projects in its 2004 *Energy Outlook* that the world will use 121 million barrels a day by 2030.[5] The United States uses about a quarter of all the world's oil, with 5 million of our 20 million barrels per day coming each year from the Middle East, where most of the world's oil reserves are found. As Amory Lovins has calculated, an increase of just 2.7 miles per gallon (mpg) in the U.S. vehicle fleet would entirely eliminate our current Mideast oil imports (and the need to protect and fight over them).[6]

Meanwhile, U.S. domestic production of oil continues to decline since it peaked in the 1950s. Even new sources, if found and produced, would not change this basic picture. When we will run out is debatable, but all experts agree. Oil is *finite.* It built up over millions of years of geologic time. It *will* run out. In the meantime, oil will continue to seep, ooze, leak, get spilled in accidents, and be turned into millions of tons of CO_2 (and other toxic by-products) until it is used up in our cars and homes and factories, never to return. What Paul Roberts makes hauntingly clear in the opening to *The End of Oil* is that even in oil-rich Saudi Arabia, major fields have been depleted.[7] No matter how you look at it, the production of oil and its combustion harms human health, the environment, and is the leading contributor to global climate change. It will begin to run out sometime this century, while you and I are still alive.

Natural Gas

The same is true for natural gas, the substance that comes out of your stove with the familiar odor added to it so you remember to turn it off, or notice leaks which can explode. Gas is seen increasingly as a substitute or supplement to oil. Natural gas is also the result of eons of geologic processes. But it, too, is finite. Gas is less environmentally damaging than oil; it can be burned more efficiently with less CO_2 per unit of energy produced. That's why you see some city buses or company vans or other vehicles proudly proclaiming that "This vehicle is running on clean natural gas." It should say "cleaner" and "we'll figure out the greenhouse gas later." But even though it does cause global warming, gas can have immediate health benefits, which must be taken into account when we are faced with hard choices. I found that out when I worked on a project concerned with the health of *lladreros,* or brick makers, in Ciudad Juárez along the U.S.-Mexico border, just across the bridge over the Rio Grande from El Paso, Texas. The brick makers build small, rounded, adobe kilns in their neighborhoods. Then they burn almost anything flammable—from scrap wood treated with preservative chemicals to old tires—to heat their ovens. The result, depending on prevailing winds, is serious air pollution for everyone in Juárez and El Paso. But it is most serious for the workers themselves, their families, and their neighbors. They suffer inordinate rates of asthma and other ailments from the resulting black haze. The PSR project worked to alert the community and local health providers and to engage citizens on both sides of the Rio Grande in advocacy for cleaner air. But we also welcomed the interventions of the Sandia National Laboratory, which helped design more efficiently burning furnaces, and El Paso Natural Gas, which assisted communities in switching to gas, a much more efficient, cleaner fuel.[8] The benefits were immediate. At present, especially in the developing world, natural gas is far superior to burning coal or older biomass fuels like wood or dung. The air is cleaner, health effects are less severe, and there is no trail of environmental damage similar to the coal mines that scar and pollute the landscape around the world. In fact, many countries hope to meet their Kyoto carbon reduction targets by switching from coal to natural gas. Gas is also a cheap feedstock for producing hydrogen for future fuel use.[9]

Nevertheless, in addition to CO_2 emissions, the most serious obstacle to natural gas may be its vulnerability to terrorism or accident. In the coming years, gas will have to come to the United States either from Canada via vulnerable pipelines or from tanker ships in the form of liquefied natural gas (LNG). It must be stored in facilities that are highly flammable, explosive, and an easy target for terrorists, foreign or domestic. There are already four LNG terminals in the United States, with thirty additional ones proposed.

As critics like Representative Ed Markey (D-MA), who opposes an LNG facility in Boston Harbor, point out, there are still no serious plans to prevent terrorist attacks on either tankers or terminals. A single LNG tanker, about three football fields long, could, if ignited, whether by accident or attack, destroy an entire city like Boston. That would take out, along with everything else, Faneuil Hall, the Fine Arts Museum, and Fenway Park, creating a fiery and frighteningly real curse of the Bambino.[10]

As Julian Darley makes clear in *High Noon for Natural Gas: The New Energy Crisis,* natural gas, too, will need to be phased out and replaced by renewable fuels and by energy efficiency—the sooner the better. Like oil, the production of natural gas in the world is about to peak. Declines in production and, ultimately, its end will follow. Already, producers are looking to the Alberta tar sands to produce more gas. But this will prove to be highly expensive and environmentally destructive. In both 2001 and 2002, more natural gas was used than was discovered. In September 2003, a little-noticed report from the National Petroleum Council (NPC) concluded after an eighteen-month exhaustive study of North American gas supply and demand that, in Darley's words: "the Titanic is about to hit the iceberg."[11] The decline will affect far more than just electricity production; natural gas is used to help produce a range of products from fertilizers to fabrics. The world cannot immediately shut down and turn off all natural gas. A delicate strategy is needed in which renewable energy is pushed very hard—as it is in organic farming and sustainable production techniques—while natural gas is phased out. But its risks should be noted clearly. It is nonrenewable. It gives off CO_2 emissions that will increase global climate change, though more slowly than coal. And it is seriously vulnerable to sabotage—from wellhead to pipelines to tankers to LNG terminals.

Nuclear Power

Nuclear power has a long history inextricable from nuclear weapons. In its early days, nuclear power became part of Cold War political and cultural battles. The United States did not want to leave assisting the energy needs of developing nations to the Soviets. The Atoms for Peace program developed by the Eisenhower administration made nuclear power available to other nations. But it also served to divide and blunt growing global and domestic opposition to nuclear weapons testing in the atmosphere. Nuclear power gave the gloss of hope and "energy too cheap to meter" to all things nuclear.[12] Finally, as a carrot to get nations to give up or forswear nuclear weapons, the Nuclear Non-Proliferation Treaty (NPT), signed by Lyndon Johnson in 1968 and extended permanently in 1995 by Bill Clinton, guaranteed signatories access to nuclear power. Under article IV of the treaty, nations that sign away their right to nuclear weapons are helped to get nuclear power. All that is required is that they agree to international inspections. These are to assure that fuel is not being diverted to nuclear weapons production.[13]

The result has been that some experts support the NPT, despite its pro-nuclear article IV, while still opposing nuclear power. They would like to see article IV changed, but not at the risk of unrestrained nuclear weapons proliferation. Other experts support nuclear power *and* the NPT. They believe adequate safeguards against bomb proliferation can be maintained. Nuclear power supporters have tended to be technological optimists and a bit scornful of those, especially nonphysicists and citizens, who oppose or fear nuclear power. But for global warming, all this renewed interest in and debate over nuclear power might have remained a somewhat minor, academic debate. Now as climate change has begun to get more attention, various prestigious reports have been pushing new nuclear power plants. One of the most influential is a 2003 MIT report, *The Future of Nuclear Power: An Interdisciplinary MIT Study*.[14] Its lead author is John Deutch, former head of the Department of Energy and the CIA. Another prominent advocate in the pro-nuclear camp is Richard Garwin. He is an Enrico Fermi Award–winning physicist with tremendous respect in the arms control community and on Capitol Hill. He has cowritten *Megawatts and Megatons: A Turning Point in the Nuclear Age?* with the French Nobelist Georges Charpak. It lays out the case for using nuclear power to meet the world's growing energy demands, while warning against the dangers of nuclear weapons.[15]

On the surface, the arguments from Deutch and the MIT authors and from Garwin (who disagrees with Deutch on nuclear bomb testing) seem sensible. They represent some of the key arguments generally used by proponents. These include the following: that energy efficiency and renewables will not be sufficiently developed by midcentury to meet the world's growing energy demand; that nuclear power is carbon neutral, as nuclear power plants themselves give off no CO_2 emissions (though the industrial processes used to mine, enrich, shape, and transport uranium certainly do); that the risk of catastrophic accidents is extremely low; and that even if human exposure to any amount of radiation is harmful and can lead to cancer, exposure throughout the entire nuclear process is minimal and acceptable. Nuclear power advocates further assert that nuclear wastes can be kept away from human exposure for the eons needed to prevent risk. They argue that rigorous international inspections can prevent nuclear arms proliferation, and that nuclear plants and their fuel rod pools can be adequately safeguarded from terrorist attack. Finally, especially since Kyoto, they emphasize the catastrophe awaiting humanity from climate change if global power needs are not met with a mix of fuels, including nuclear power. To seal the deal, they generally compare nuclear power plants directly to highly polluting coal-fired power plants.

But to assess the risks and benefits of nuclear power, it is necessary to look at the entire life cycle of the nuclear fissile materials—uranium and plutonium. They do not exist in a usable state in nature. Uranium, the primary fuel for nuclear power reactors, must first be mined and milled. Then it goes through an elaborate refinement or "enrichment" process that uses large amounts of energy. Plutonium-239, which can be used for fuel or bombs, can only be produced as a by-product of nuclear power reactors fueled by uranium. Both enriched uranium and plutonium are highly radioactive and, once created, last extremely long times. They stay radioactive based on their half-life, which is the time it takes for half of either material to break down into lighter radioactive elements, and ultimately into harmless nonradioactive ones. Plutonium-239—one of the most lethal substances on earth—was first produced in quantity at the Hanford Nuclear Reservation in Washington State for the bomb that destroyed Nagasaki, Japan. It has a half-life of 24,400 years. It remains dangerous for about a quarter of a million years—longer than the entire time in which modern humans evolved and about twenty times as long as all of settled human civilization since the Ice Age. The half-life of uranium-235, or highly enriched uranium (HEU), the usual choice for reactors and bombs, is even longer—an incredible 704 million years. Thus, radioactive fuel rods from a power plant remain radioactive and dangerous far beyond any reasonable regulatory period—almost beyond imagination.[16]

Uranium—the bomb material that destroyed Hiroshima and has been contained in all of the more than sixty thousand nuclear weapons produced by the United States since that time—has been mined in the United States mainly on Navajo territory. It has led to substantial health problems for the Navajo nation, or Dine people.[17] Dine legends tell of how they emerged in prehistoric times into the present world and were offered the choice of two yellow powders. They chose corn pollen. The other possibility was what we now call yellow cake or uranium oxide—a yellow dust that has fueled the nuclear age. It is abundant in the American Southwest. The Dine spirits warned that the yellow dust should be left alone. But from the late 1940s through the mid-'80s it was mined by pick and shovel and blasted and hauled across the West, ending in mountainous piles. Thousands of Navajos worked in these mines. More than a thousand abandoned shafts are still on Navajo land that ranges across parts of New Mexico, Arizona, and Utah. All in all, more than five hundred uranium miners died of lung cancer between 1950 and 1990, and hundreds more will die according to a Public Health Service study. Since 1990, under the Radiation Exposure Compensation Act of 1990 (which Physicians for Social Responsibility actively supported) Americans exposed to radiation through uranium mining and milling or through nuclear weapons testing are eligible for compensation. John Fogarty, head of the Indian Health Service in New Mexico, and a PSR board member, works at the Radiation Exposure Screening Education Program. It is one of four clinics in the Southwest that have screened between three thousand and four thousand miners. The Navajo miners run a risk of developing lung cancer that is twenty-eight times greater than other Navajos not exposed to uranium and its by-products, called "radon daughters," which adhere to dust particles and are inhaled. Fogarty is concerned that mining is planned to resume. Promoted by Hydro Resources, it would involve pumping water, bicarbonate of soda, and dissolved oxygen to dissolve uranium and bring it to the surface for drying and processing. This in situ process, according to Fogarty and the Southwest Information Center, could lead to contamination of the Westwater Canyon aquifer and ultimately to kidney disease. Once above ground and dried, uranium powder remains a risk to anyone who inhales even small quantities of dust.[18]

Further health risks, wastes, and energy consumption are associated with the refining or enriching of uranium. So is nuclear weapons proliferation. In *Insurmountable Risks: The Dangers of Using Nuclear Power to Combat Global Climate Change,* author Brice Smith of the independent Institute for Energy and Environmental Research points out that just 1 percent of the expanded enrichment capacity needed for the growth scenario projected from the MIT report would be enough for 175 to 310 nuclear weapons per year. Although President Bush and others have proposed ingenious plans to avert these risks—such as an International Fuel Bank where states could purchase enriched uranium without building enrichment plants—none is foolproof.[19]

Arcane debates over health risks and scenarios for future proliferation reflect what is at stake with the entire nuclear cycle. Nuclear power and nuclear weapons cannot truly be disentangled. This was recognized as early as 1946 by noted atomic scientists like Robert Oppenheimer. Take nuclear wastes, for example. There have been extensive battles over the Yucca Mountain Nuclear Repository. The potential health risks from transportation, burial, leakage, and the rest are quite real. They will remain so for millennia. But they also generate more passion than one would ordinarily expect for a pollution issue. What has been at stake for pro-nuclear power and pro-nuclear weapons advocates in the Yucca debate is an entire industry. It supports both nuclear power and nuclear weapons. Like a horse and carriage, you can't have one without the other.

For those advocates who believe that the United States needs to maintain, and ultimately upgrade, its nuclear weapons arsenal, new bomb making facilities will need to be built. Public fears of radiation, whether from wastes or bombs must once again be assuaged and overcome. Yucca Mountain is meant to receive spent fuel rods and other high-level nuclear wastes from *civilian* nuclear reactors. It is almost always described this way. But it is also slated to receive *military* high-level wastes from the production of U.S. nuclear weapons, such as the vast amounts of radioactive materials at the Hanford Nuclear Reservation in Washington State.[20]

Because the pools that hold spent nuclear fuel rods at American power plants are full, no more can be opened and others will need to be shut down unless Yucca becomes operational. Similarly, unless nuclear wastes from the decades of U.S. nuclear weapons production during the Cold War can be disposed of safely, public acceptance of the *new* nuclear weapons facilities being planned will remain unacceptably low. As opposition to burying nuclear wastes has remained strong, it has led to pressure to use or build nuclear plants to reprocess (recycle or transform into safer wastes) nuclear fuels and weapons-grade fissile materials from Russia. Given the danger of stimulating a world plutonium economy that would risk nuclear bomb proliferation, the Carter and Ford administrations ruled out such reprocessing facilities for the United States. Now that they are being used, however, a number of safeguards have been implemented. But reprocessing generates another set of nuclear wastes and problems. And because plutonium is always present and can be recovered or stolen, risks will multiply if reprocessing expands. Currently, only a few such plants operate, but the Thorp plant in the United Kingdom has already reported that it cannot account for forty-nine kilograms of plutonium, enough for six nuclear bombs, while the Japanese have reported having lost enough plutonium for thirty-four nuclear weapons.[21]

All of these dangers are considered manageable by pro-nuclear advocates. And the health risks involved often get particular disregard or are considered trivial. Richard Garwin even goes so far as to compare the relatively small number of estimated deaths and illnesses directly attributable to nuclear power (not counting meltdowns, terror attacks, or bombs) with the three million people per year who die from smoking tobacco. He says: "One might perhaps, expect that Greenpeace will henceforth turn away from nuclear concerns and mobilize international brigades that will sink cigarette transport ships and set fire to tobacco fields. We absolutely oppose such violence, but three million deaths per year should be kept front-and-center in our consciousness."[22] Aside from the incorrect inference that Greenpeace supports violence (it was the French government that sank the Greenpeace ship *Rainbow Warrior* and killed Greenpeace nonviolent anti-nuclear campaigners),[23] and the difficulties of measuring health effects through epidemiological studies of radiation, Garwin fails to appreciate one of public health's primary tenets. As their first principle, physicians and public health scientists seek *to prevent* human harm. As primary prevention, they would seek to *eliminate* the building of nuclear weapons, nuclear power plants, *and* the cigarette industry. Each has and can cause catastrophic harm to humans. When *primary* prevention is not possible (usually because of powerfully entrenched economic and political interests), then *secondary* prevention (reducing risk) and *tertiary* prevention (offering antidotes and treatment) are brought to bear. It is *not* a public health approach to suggest that nuclear power may risk catastrophe, but we will nonetheless calmly accept it and proceed.[24] Large numbers of preventable deaths may occur first, but we will at some point find solutions to this problem. Unfortunately, solutions are not at hand. And worse, in order to reduce carbon emissions very much, a rapid and large expansion of nuclear power reactors will have to occur without critical health and safety problems being resolved.

In *Insurmountable Risks,* Brice Smith meticulously reviewed estimates in the 2003 MIT report. He calculates that even the huge, unprecedented growth anticipated would still leave U.S. carbon emissions climbing rapidly. MIT posits a growth scenario for nukes to meet projected electricity needs in 2050 of 1,000 gigawatts (GW). Because all of the reactors in operation today will need to be shut down by 2050 (the materials in the plant become brittle and unsafe because of exposure to radioactivity), Smith estimates that 2,300 GW of capacity will actually need to be built. In the United States, under the MIT scenario, that means putting up three hundred new reactors, or one every fifty days. This astounding rate, which the MIT authors admit is unprecedented, would still only raise the portion of world energy from nuclear power from 16.3 percent in 2000 to 19.2 percent in 2050. In the United States, carbon emissions—even with the nuclear construction envisioned in the MIT report—would grow from 13 percent to 62 percent over year 2000 levels. IEER therefore projected a more realistic scenario in which carbon emissions are held at 2000 levels through *even more* nuclear plant capacity. To do that, the world will need between 1,900 and 3,000 GW—much of it in the United States.[25] In short, the MIT analysts base their advocacy of nuclear power as an antidote to greenhouse gas emissions from coal on seriously unrealistic estimates of what would be needed and the risks entailed.

Take a nuclear accident as an example. Garwin, who has been closely associated with nuclear power technologies, was called in during the Three Mile Island (TMI) accident in 1979 to find a solution to the problem of a large hydrogen bubble that had formed. Hydrogen is explosive when mixed with oxygen, and the possibility of a hydrogen explosion was considered very possible and serious at TMI. Garwin and other specialists worked frantically to find ways to avoid such a calamity. Ultimately, oxygen, that would have made an explosive mixture, was not produced. But it could have been. As Garwin himself states: "*The formation of the hydrogen bubble had not been foreseen in any of the accident scenarios imagined by the governmental regulatory authorities*" (emphasis in the original).[26]

The safety and reliability of U.S. nuclear power plants is regulated by the Nuclear Regulatory Commission (NRC). It is a descendant of the Atomic Energy Commission (AEC), whose formation was considered a victory by liberals in the 1940s when nuclear power might have been controlled by the military. Nevertheless, since early Cold War days, the AEC has been an advocate for, as well as a regulator of, the nuclear industry. The AEC, now called the NRC, is independent from the Energy Department and the nuclear industry. But it "does not require that nuclear plants withstand all conceivable threats." The NRC did not, for example, consider the possibility of truck bombs or explosives destroying containment vessels until after the 1995 Oklahoma City bombing. And even after 9/11, security procedures at nuclear power plants have been found consistently to be inadequate for even a relatively simple terrorist attack. If the most severe accidents or attacks are unlikely, then in the eyes of the NRC, nuclear power plants are worth the risk. Right now the NRC's primary requirement in the face of catastrophe is that "plants be able to endure any earthquakes foreseeable for the site." In a word, the 104 nuclear power plants in operation in the United States, and those likely to come on line if the current push for nuclear power accelerates, still run the risk of either a terrorist attack or a core meltdown. "Can the public be confident that there will be no more accidents like Three Mile Island?" asks Garwin. "More needs to be done."[27] The exact figure calculated by Garwin for the 103 U.S. reactors active when his book was written is one reactor meltdown over the next one hundred years.[28] Obviously, if more reactors are built, the risk will increase (as will the fatalities and casualties as U.S. population continues to grow and the nation further urbanizes.)

I remember to this day the effect of Three Mile Island (TMI) on my American studies students at Temple University in Philadelphia. Temple is about 110 miles east of Harrisburg, where TMI still stands along the Susquehanna River. On March 29, 1979, after a coolant leak and power surge related to operator error, the nuclear core of TMI was exposed and began a near meltdown. Had it occurred, radioactive fallout would have spewed eastward toward my classroom and my home. For the 20 million of us who lived in the potential radioactive fallout plume, the news was especially chilling. As TMI's core was partially melting, *The China Syndrome,* an anti–nuclear power movie with Jack Lemmon and Jane Fonda, was playing in local theaters. Long lines immediately formed, and I watched moviegoers eagerly take and begin reading brochures, fact sheets, and petitions thrust at them by local anti-nuclear activists. The next day in class, my students *demanded* to talk about the news of TMI, their fears, the facts, and the social movements that could prevent such a catastrophe. TMI demonstrated unforgettably that worst-case scenarios can happen.

As if to prove the point, a full meltdown did occur in 1986 at a far less modern nuclear power plant in Chernobyl, Russia. Experts still argue over the extent, nature, and lasting consequences of that huge release of radiation over Europe (the prevailing winds blew northwest from Kiev, Ukraine). But large numbers of Ukrainians, Russians, and other Europeans did die. Many more will ultimately develop cancer, suffer, or die. When Unit Four of the Chernobyl nuclear power plant in the Soviet Ukraine exploded from steam pressures following an episode of runaway power, the reactor burned for ten days. Leakage of radiation continued far longer. Thirty-one people were killed fairly soon from acute radiation sickness as a result of severe exposures, 130,000 people were evacuated from the area, and high-fallout areas extended from 60 to 125 miles from the accident. Two months later, another 113 villages were evacuated. Finally, more than 220,000 people were forced out of their homes.[29]

To understand just how devastating the meltdown at Chernobyl was, we need first to recognize the uncertainties and difficulties associated with estimating deaths through epidemiological studies. Estimates range from extremely understated ones that choose to look for best cases and minimal dangers, as did the scientists at the NRC. This was certainly the initial Soviet approach. The Russians downplayed the accident. Then they consistently failed to include continuing exposure to radiation that had been absorbed into food, causing additional risk to humans. In areas with substantial food production in the Soviet Union, for example, the Soviet focus on exposures to external radiation (and not internal exposures from irradiated food) led to an underestimate of actual exposure to cesium-137 by more than two-thirds. The Soviet estimates also neglected exposures to radioactive iodine-131, which humans absorb from cow's and goat's milk. The sale of milk around the Chernobyl site was banned for more than a year after the meltdown and affected between 20 million and 25 million people. Nevertheless, milk was still produced and consumed throughout the Chernobyl area. Iodine-131 also affected people far outside the Soviet Union, with exposures in Italy, Sweden, Germany, the United Kingdom, Hungary, Poland, Czechoslovakia, Austria, Switzerland, and Romania beyond the level of 15,000 pico curies that requires action.[30]

An early 1988 estimate of 17,400 excess thyroid cancer deaths just from iodine-131 alone was done by the Lawrence Livermore Laboratory, the Electric Power Research Institute, and the University of California–Davis. Sixty-two percent of the deaths occurred outside the Soviet Union, including 3 percent in North America. This study also found an "upper bound" high-end possibility of as many as 51,000 excess cancer deaths. In 1987, the EPA estimated 14,000 cancer deaths. A UN study in 2005 concluded that some 4,000 cancer deaths were likely. This estimate, however, was based on populations exposed to high levels of radiation. It left out lower-dose exposures. Thus, cancer deaths from Chernobyl are likely to range somewhere between 15,000 and 50,000 fatalities.[31]

Although no full meltdown has yet to occur at a U.S. nuclear power plant, we can estimate from Chernobyl, Hiroshima, Nagasaki, and other human radiation exposures what the risks from an accident are. A 1981 study done by the Sandia National Laboratory estimated the peak *early* fatalities from each of the major U.S. nuclear power plants. They also estimated the number of cancer deaths likely after a longer period of time. This startling government report was only released to the public after concerted effort by the Union of Concerned Scientists. By early fatalities, Sandia means only those occurring during the first *nine years* following a meltdown. We know, however, that many of the radioactive elements released will remain dangerous far, far longer. Here are some of the U.S. government estimates of just *early* fatalities.

The Salem 1 and 2 nuclear power plants along the Delaware River in Wilmington, Delaware, lead the list with one hundred thousand early deaths, followed by an additional forty thousand cancer deaths. In each case listed, the estimates assume prevailing wind patterns, though these can shift. In the case of Salem, Philadelphia residents know that the estimates are higher than some cities because any fallout would likely blow right over the City of Brotherly Love. The same is true for the Limerick 1 and 2 reactors to the west of Philadelphia. A meltdown there, according to the Sandia labs, based on 1980s population figures, would cause seventy-four thousand deaths, followed by thirty-seven thousand more fatal cancers. Other cities are high on the list as well, even if the nearby reactor is far from the city limits. At the "low" end is the Pilgrim plant in Plymouth, Massachusetts. It would sustain only twenty-three thousand fatal cancers over time. This assessment presumes that radioactive debris would be blown out mostly over the ocean, not Fenway Park. The estimates for New York's Indian Point 2 and 3 nuclear power plant assume fifty thousand deaths.[32] But even this statistic, equal to about twenty-seven attacks on the World Trade Center, fails to account for deaths associated with any major catastrophe from panic and flight, especially on the crowded roads in the New York City metropolitan area.

In looking at nuclear catastrophes, we should remember that these estimates (and public fears) of death from radiation tend to focus only on cancer deaths. But the most recent National Academy of Sciences report on the biological effects of radiation, the 2005 BEIR VII report, points out that exposure to high-level radiation, as in a nuclear accident, also causes other life-threatening conditions. These include stroke, heart disease, and diseases of the digestive, respiratory, and blood cell systems. The BEIR VII report also states that "there are extensive data on radiation-induced transmissible mutations in mice and other organisms." It concludes from this that "there is therefore no reason to believe that humans would be immune to this sort of [inheritable genetic] harm."[33]

But perhaps the most telling argument against nuclear power as a solution for global warming lies in its security risks. Immediately after 9/11, it became clear that whatever their carbon emissions or wastes, nuclear power plants are sitting time bombs. Flight 93, which crashed into a field in Pennsylvania when United Airlines passengers and members of the flight crew tried to regain control from the hijackers, was headed toward the nation's capital. But it might just as easily have aimed for and destroyed TMI or any other nuclear power plant in Pennsylvania with far more devastating results.

As I lobbied on the energy bill with other CEOs from the Green Group, only two of us were in a position to seriously raise these risks. It was not because my colleagues were timid or trapped inside the Beltway. The mission of their organizations simply did not include assessing the risks of terrorism and war. And so it fell generally to me and to Brent Blackwelder, the president of Friends of the Earth, to try to convince various senators to cut the subsidies for nuclear energy. Blackwelder—an unassuming, scholarly looking PhD with the passion of a radical environmentalist and the straightforward ethics of his descent from Episcopalian ministers—had already dug out of his files a copy of a 1982 report from the Federal Emergency Management Agency (FEMA). It unmistakably warned about the vulnerability of nuclear plants to sabotage and attack. PSR issued fact sheets and press releases about these risks and joined coalitions in New York that once again called for the shutdown of the Indian Point power plant northwest of the city. If blown up or destroyed, Indian Point is within range of some 50 million people. Renewed and intense concern about the safety of nuclear power followed—from congressional hearings, from defense experts like Senator Sam Nunn, and from a voluminous series of books, articles, pamphlets, and press releases from the remaining anti-nuclear groups. But they produced little change. They could not penetrate the saturation media coverage of the president and his constant assurances that all necessary security procedures were being put rapidly in place.[34]

Although President Bush himself once cited the threat of terrorist attacks on a nuclear power plant, the danger was soon downplayed and ultimately neglected. I personally witnessed the difficulty when I met directly with the head of the Nuclear Regulatory Commission (NRC) and his senior scientists. We had come to discuss a computer-based analysis that PSR had carefully constructed of an attack on the fuel rod pond at Indian Point, New York.

I laid out elaborate computer printouts and estimates of explosions, dispersals of radioactive materials, and the like in front of a small group at NRC headquarters in Rockville, Maryland. I was joined by Ed Lyman, a nuclear power expert, now with the Union of Concerned Scientists, and a couple of others. We debated whether a small Cessna private plane packed with explosives could reach, dive into, and destroy the holding pond, leading to the ignition of fuel rods and a massive radioactive release. If so, where it would spread? Were evacuation plans and other civil and medical preparations adequate? I grew up on Long Island and lived through many a holiday or weekend traffic nightmare, even after the Long Island Expressway was built. So I nearly lost patience as NRC's chief scientist kept explaining how, even if such an attack were possible, people could be evacuated. Yet the Shoreham nuclear power plant, only miles from my parents' home, had finally been prevented from opening by local activists. They had been able to prove in court, in the event of a meltdown, that no credible scenario could be found to evacuate the residents of Long Island. Nevertheless, the NRC staff refused to even consider the possibility that we might be right about the ability of a small plane to destroy the fuel rod ponds or the impossibility of real evacuation. They remained rigid in the face of contrary expert analysis and even after American forces had found blueprints of nuclear power plants in the bunkers of al-Qaida.

As of 2008, no serious planning or preparation to protect American nuclear plants has yet to be undertaken; none have been shut down or dismantled in the light of the potential for terrorist attacks.[35] The dangers and the medical effects have been reviewed in a PSR report, *The U.S. and Nuclear Terrorism: Still Dangerously Unprepared.* It addresses the threats of nuclear terrorism from a nuclear bomb in New York City, a radioactive, conventional "dirty" bomb in Washington, D.C., and an attack on a nuclear power plant. The nuclear power scenario focuses on the Braidwood nuclear reactor sixty miles southwest of Chicago. It is one of thirteen nuclear power plants operating in northern Illinois and produces 2,500 megawatts (MW) of electricity. Its core of nuclear fuel rods contains twenty to forty times the amount of radioactive material released in a small nuclear bomb explosion. Using the same sophisticated Department of Defense threat assessment software as in previous analyses—along with enhanced population, medical, and other data—PSR concluded that if an attack breached the containment dome, "the resulting plume of radioactive materials would extend north from the reactor itself to the northern edges of metropolitan Chicago, and east into Indiana and Michigan. . . . Over 7.5 million people would be exposed to radiation over the maximum allowed annual dose with 4.6 million receiving the equivalent of the maximum allowable occupational exposure for one year. More than 200,000 would receive high enough doses to develop radiation sickness and 20,000 might receive a lethal dose." In rather clinical prose, the report explains that medical facilities, including all 113 hospitals in the area, would be overwhelmed. Twenty thousand practicing physicians would be exposed to more than allowable occupational levels, as would firefighters and other emergency responders. All told, within a year, some 44,000 people would die. More than a half million citizens would eventually perish, while economic losses in the greater Chicago area could top $2 *trillion.*[36]

Nuclear power plants will always present serious health and security risks. These are inherent in an industrial process that uses radioactive materials that far outlast human lifetimes and from which bombs can be made. And we must not forget the health problems, environmental pollution, and vast energy use throughout the nuclear production cycle—from mining, milling, enrichment, and transportation to ultimate disposal or reprocessing. Finally, despite assurances to the contrary, nuclear power can never be disassociated from terrorist attacks. As we have seen from the domestic terror bombings at Oklahoma City or anthrax attacks on the Capitol, such risks are not limited to international enemies. And, in an age of missiles, as we have seen in the 2006 Mideast war, areas long considered invulnerable can some day come under attack.

COAL

If nuclear power has gotten renewed interest thanks to America's steadily rising electricity demands and concern about global climate change, so, oddly enough, has coal, the dirtiest conventional fuel. Worldwide, given current trends, according to the International Energy Agency, we can expect by 2030 to see some 1,400 gigawatts (GW) of new coal-fired power plants. If emissions from all these new plants are not controlled, CO_2 in the atmosphere will increase by 7.6 billion metric tons—about half of all the carbon emissions during the entire Industrial Age.[37] Here in the United States, about 145 GW of new coal-fired electricity is projected by 2030, giving off 790 million metric tons of CO_2 per year. Meanwhile, utilities have already announced plans to build 151 new coal-fired plants with 90 GW of power. As most power plants built in the past fifteen years were powered by cleaner, more efficient natural gas and only eleven GW of coal plants were built between 1991 and 2003, we are clearly witnessing a coal renaissance.[38] There are three main reasons. First, coal is an abundant, relatively cheap fuel found in many parts of the United States. It is unlikely to run out under any scenario for at least a couple of hundred years. Second, because of its low cost (not counting its human health and environmental costs), and widespread availability, it has developed a strong political base. Its strength was reflected in the first, early vote of 95–0 against going forward with Kyoto unless developing nations were included. It was Senator Robert Byrd of West Virginia, a leading coal state, who pushed that effort. It also had the support of the United Mine Workers, various utilities, and numerous businesses that count on coal for electricity.

Given this history, pragmatists in the environmental and political world have been looking for ways to attract or mollify key senators from coal states and to eliminate a major source of legislative opposition to both clean air and climate legislation. They reason, because coal is racing ahead in the developing world and in the United States, it is only prudent to try to reduce its impact on global climate change. Although I do not agree, these realists are not convinced that public opposition or the attractions of renewable energy are strong enough to stop coal. At a minimum, they argue, we should invest in policies that will minimize its hazards. Thus the attraction to "clean coal" technologies that would lower emissions, reduce health impacts, create new investments and construction primed by government subsidies, and increase or save jobs in unionized states.

Once Clinton-era officials and experts who witnessed the power of the coal lobby firsthand were cast out into the world of consulting, think tanks, and NGOs, carefully nuanced policy suggestions for using coal began to appear. A delicate dance began between political pragmatists and traditional environmental groups—most of which, though not all, continue to oppose coal mining and combustion under any circumstances. Reid Detchon, for example, who had been a Clinton official and Turner Foundation officer close to environmental groups, heads up the Energy Future Coalition (EFC). It includes a few pragmatic environmental groups like the Union of Concerned Scientists and NRDC, along with former Clinton officials, corporate heads, foundation officers, and representatives of the coal industry. Though NRDC and UCS placed caveats on using coal only if certain very difficult environmental conditions were met, the prestige and clout of the EFC participants and its report, *Challenge and Opportunity: Charting a New Energy Future,* set an important tone. It proposed research and development expenditures for gasified coal to be used as a fuel. Coal is gasified by heating it to 2000°F in a sealed chamber where it breaks down into various chemical components—including a kind of natural gas—that can then be used in many ways. But the process gives off as much CO_2 and other elements like mercury as burning coal. These need to be disposed of. Hence, the report also called for the exploration of safe ways to bury any CO_2 from coal technologies in deep geological formations ("carbon sequestration"). If such technological breakthroughs could be found, coal could serve as a relatively clean, low-carbon fuel during the mid-twenty-first century.[39] Other policy recommendations for researching "clean coal" and new coal technologies have since appeared in the National Energy Commission's report[40] and in energy proposals for climate change from Al Gore.[41] And, while promoting additional dirty coal plants, the Bush administration has touted future technologies as a solution through its FutureGen Project, a federal bonanza for whichever state gains the more than $1.5 billion experimental facility.[42] Given that current coal technologies are likely to go forward in the United States and in the developing world, especially in China, environmental groups are faced with a familiar dilemma.

If they want to be taken seriously in policy circles and be seen as reasonable participants in the world of business and energy dialogue, it is difficult to simply keep saying a resounding no to coal. Hence, the willingness, with strong reservations, for clean air and climate experts like David Hawkins of NRDC, an assistant EPA administrator under President Carter, to consider how to help make coal technology at least safer and cleaner. Hawkins says that "even if a grand bargain were struck with the coal industry on dealing with the downstream effects of carbon emissions, the environmental community is not going to walk away from the concerns about the upstream side, where the coal comes out of the ground." At the same time, the current uses of coal are so horridly polluting and environmentally devastating and any technological "fixes" so distant and futuristic, that to accede to discussions about using coal infuriates key grassroots activists. They are not prone to compromise. Indeed, the NRDC online magazine, *OnEarth,* got plenty of heated replies when it ran a cover story called "How to Clean Coal." Roel Hammerschlag, a respected analyst with Seattle's Institute for Lifecycle Environmental Assessment, who is quoted in the article, says simply: "Coal is the enemy. [But] we're going to run out of oil in the next century. . . . So coal will replace oil. . . . There's at least 300 years' worth of coal still in the ground. That's enough to raise atmospheric CO_2 to insanely high concentrations—10 times preindustrial levels."[43]

A key question is whether proposals to capture and bury the CO_2 from coal are really reliable and likely to be seriously implemented. Keep in mind that all that CO_2 must, like nuclear waste, disappear essentially forever. And, as with nuclear power, many of the proposed solutions are not yet technologically feasible or tested. Interest in new coal technologies and especially carbon sequestration began seriously in the 1990s. Scientists and policymakers (and those with an interest in using coal) explored two main means of removing the CO_2 emitted by coal burning from the atmosphere. Although some attention has been given to putting CO_2 directly into the deep ocean, carbon sequestration or burial in geological structures beneath land or the ocean sea bed has emerged as the leading candidate. It is important to keep in mind that even though CO_2 is an essentially nontoxic gas, there are a number of risks associated with it. In the presence of moisture, for example, CO_2 forms carbonic acid, which lowers the pH of any water that it contacts. This increase in acidity can cause lead and other trace elements to leach out in ways similar to acid mine drainage. The result is that contaminated water can infiltrate groundwater and drinking water sources and cause major health effects in humans or animals.[44] At the depths where it is likely to be stored in sequestration projects, CO_2 also increases in density to about 65 percent of water, about the same as crude oil. Thus, there is upward pressure on any cap material above it and the potential for leaks. If CO_2 leaks from underground storage, the result can lead to more than simply adding to greenhouse concentrations. Leaks have already proved to be deadly. In California in 2006, for example, members of the Mammoth Mountain Resort ski patrol were inspecting the mountain for potential avalanches when three of them fell into a volcanic fissure. Very high concentrations of CO_2 quickly killed the three ski patrollers and hospitalized four rescuers. CO_2 had eroded the snow from beneath, creating a gas chamber that the skiers' weight caused to collapse. Police tested air at the scene and found high levels of CO_2 which, though nontoxic, deprives organisms of oxygen and causes asphyxiation. In addition to this human tragedy, geologists also believe that significant areas of vegetative destruction on Mammoth Mountain are related to leaking CO_2. First noticed by U.S. Forest Service Rangers, plants on one side of the mountain are dying. Although the CO_2 is not from human sources and burying, but from shifts in magma below the surface, an estimated 530 tons of CO_2 are released into the soil every day, suffocating the roots of trees and other plants. So far, some 170 acres of land has become devoid of vegetation.[45]

Additional problems can come from CO_2 leaks. In 1986, for example, heavy rains in Cameroon triggered a mudslide into small, deep Lake Nyos formed in an extinct volcano. For centuries CO_2 had been slowly seeping into the lake from bicarbonate springs, causing CO_2-rich water to fill the lower layers of this stratified lake. When the mudslide hit, the water layers rapidly inverted. When the CO_2-laden waters from the bottom of the lake reached the surface, lower pressures turned the CO_2 back into its gaseous state. Scientists estimate that some 200,000 tons of CO_2 were instantaneously released into the surrounding ecosystem. A CO_2 cloud formed over the surrounding valleys, killing 1,700 people and thousands of animals.

A similar seismic release in Cameroon's Lake Monoun killed 37 people in 1984, while in 1979, an explosion at the Dieng volcano in Indonesia gave off 200,000 tons of CO_2, smothering 142 people on the plain below.[46]

But many scientists argue that CO_2 can be safely stored in geological deposits and that the risks of leaks are comparatively small when weighed against global warming. The process of carbon capture and storage (CCS) is being implemented in many places where CO_2 is pumped into oil fields, apparently without harm. Nevertheless, there has not been much rigorous study of such carbon burial, whether in ocean beds or geological formations on land. A strong note of caution was sounded, however, when a U.S. government study carried out in 2004 was published in the July 2006 edition of *Geology*. The experiment was led by Yousif Kharaka, a geochemist. She and her team reported that after 1,600 tons of liquefied CO_2 were injected into a depleted oil field in Texas, the underground minerals were dissolved because the CO_2 changed their acidity. The danger? Leakage into groundwater and the release of CO_2 back into the atmosphere, aggravating the greenhouse effect.[47]

Again, like nuclear power, what seem to be very small risks from carbon sequestration must be multiplied many times over, as current plans envision very large increases in coal. Even as environmental and health advocates have been winning battles against the pollutants coming from coal-fired electrical plants, more are on the way. The National Energy Technology Laboratory anticipates that up to 309 new 500-MW coal plants will be built in the United States by 2030. That could triple air pollution and carbon emissions from coal-powered generating plants. And by the end of 2006, 153 new coal plants were already proposed, with enough capacity to serve some 93 million homes. Of these, only twenty-four have even expressed any intention to use gasification technology. And some of these coal projects, like ones being developed by Peabody Coal and LS Power, are not to add to current utility capacity for local needs. They are purely speculative—being planned for sale to the highest bidder as demands for electricity rise. In the words of Carrie LaSeur, an attorney who has helped organize citizens in Waterloo, Iowa, to oppose a plant proposed by New Jersey's LS Power: "Developers are gambling that the need for power and the low price of western coal will make them very rich."[48]

Amid corn fields and, ironically, in a state that ranks with Texas at the top of renewable wind power generation, LS Power seems to have been drawn to Iowa by overtures from the Waterloo economic development agency and union endorsements for the coal plant. The area has some of the state's highest poverty rates, including East Waterloo, which has a large African American population and high rates of asthma. The LS Power proposal, which initially indicated that power would be sold to Illinois, has sparked a coalition that unites both environmental justice advocates and clean air activists. It also includes farmers like Don and Linda Shatzer whose farm features soy bean fields, an organic garden, and a pond and beach. With other citizens the Shatzers have gathered three thousand signatures against the proposed 750-MW pulverized coal facility to be constructed by 2011. Local opponents concerned about air pollution, health, and the destruction of farmland are linked to nationwide opponents; the coal that would power Waterloo comes from destructive mining practices from mountaintop removal in traditional coal centers like West Virginia and from newer, huge, "supermines" in western states like North Dakota.[49] As Craig Canine put it in the NRDC's *OnEarth* magazine, coal mining has moved in the past three decades from the "iconic" shaft mines of Pennsylvania and Appalachia with black-faced, pick-wielding men to enormous surface mines in the West where "a relatively few laborers operate the largest machines on earth such as the dragline excavators I saw working in the lignite beds of North Dakota." Most of these supermines are in the Powder River Basin and produce some 40 percent of all U.S. coal—about 400 million tons a year. Another 15 percent of U.S. coal or about 145 tons per year, comes from six hundred Appalachian strip mines, including those that remove entire mountaintops, causing "grotesque and permanent damage." When mountaintops are removed, "draglines, dozers, and huge dump trucks blast and scrape off summits and push the displaced earth into the valleys below. The procedure creates an eerily unrelieved, amputated landscape, filled with muddy stumps, acid mine runoff, and piles of toxic coal sludge."[50]

Such destruction of the environment, linked to damage to humans and ecosystems from air pollution, mercury, acid rain, and more, has led many environmentalists to oppose coal outright, like nuclear power. But as dirtier, more damaging plans move forward, the call for cleaner coal technologies as a fallback or interim measure continues. We are at a watershed moment. As Canine notes: "the largest coal plants under consideration to replace aging ones built in the 1950s and 60s will cost around $1billion, generate 1,000 MW of electricity that can supply 900,000 homes." With a life expectancy of sixty years or longer, for every year that it operates, such a plant, "will spout six million tons of CO_2 into the atmosphere—about the same as two million cars."[51]

Clearly, citizens and environmentalists must continue to oppose current coal-fired plants and any new ones being built with technologies that still release pollutants and CO_2. Realists should be aware that all funding these days is close to a zero-sum game. Support for coal diverts resources from subsidizing renewable technologies. Developing gas liquefaction plants, synthesis gas, and carrying out massive CO_2 burial will be enormously expensive. Carbon sequestration is not without serious risks of CO_2 escape over the long term, adverse health effects, and environmental damage from continued coal mining. And because it is so expensive, the industry continues to avoid it, building deadly old-style facilities and demanding financial incentives. As Jeff Goodell puts it in *Big Coal:* "sequestration is a little more complicated than just drilling a hole in the ground and pumping in the CO_2. . . . The CO_2 first has to be captured from power plants, then compressed in a supercritical fluid to be injected underground. . . . The cost of retrofitting the plant would be so high that most power companies would simply shut it down."[52] Environmentalists may want to follow a classic inside/outside strategy—calling loudly at the grassroots for "No More Coal!" while working in Washington to demand the best possible technologies as coal continues to be used. But because renewable wind power is already cost competitive with coal and has none of the severe environmental and health impacts, it may be, that as the Washington political climate improves, a concerted, united push for subsidies, tax breaks, and other federal support for the wind and solar industries may make the best sense.

Such dilemmas point to a way out that can unite environmentalists, policymakers, and local activists alike. You and I are going to have to become far more energy efficient in our own lives and demand the same of industries and public policy. We can attack the demand side of our huge appetite for power and electricity. Then we must move rapidly to renewable fuels—wind, solar, biomass and biofuels, even tidal power—that can generate the energy we need without adverse health effects, pollution, security risks, and mountains of waste left for future generations.

The disagreements among policymakers and proponents of efficiency and renewables have come mainly from the lingering notions that renewable technologies are far off, too expensive, not robust enough, and that too rapid a transition to them is not yet politically feasible. That's where you and I come in. What we say and do about new, clean, and healthy energy sources with no greenhouse gas emissions or toxic wastes can make all the difference.

8

Energy Futures

We saw in the last chapter that most of our energy comes from fossil fuels and some nuclear power. We also use it pretty inefficiently. Only a small portion is produced from clean, renewable sources. But if we need to predict the future to comprehend global climate change, we must to do the same for energy efficiency and renewables. The question is how fast can we change our energy system and how far can we go with nonfossil fuels? In order to make a prognosis and prescribe treatment, we need to be sure the prevention plan is feasible. One of the best prognosticators for renewable energy is Chris Flavin, president of the Worldwatch Institute, a nonprofit environmental think tank in Washington, D.C. Flavin is sometimes described as "an energy optimist."[1] Yet he is cogent, thorough, a true expert. He showed my class at American University a series of fascinating bar graphs with steeply rising steps. They pictured upward trends in sales and use over the past few years of renewable energy. Renewables may be a small part of the current energy market. But it is the trends and their speed that count.

The steep upward curve in renewables means we are approaching a tipping point—much like atmospheric carbon concentrations. We have had such historic energy transitions before: from wood to coal to petroleum. They can come very rapidly. More and more Americans are aware that we are nearing a dangerous tipping point for climate change. Few, however, understand that the opposite picture of that approaching doom, a sort of film negative, is emerging. Technologies, markets, and trends for renewable energy indicate, if we push for it, that we are about to move into a new energy era.

The United States went from the horse-and-buggy age, with kerosene and gas lamps, and bumpy, often dirt roads to a modern society based on oil within three decades. That's how fast some 26,700,000 cars, burgeoning cities, and a network of national highways appeared after the first major gusher of oil was struck in 1901 at Spindletop, Texas.[2] That energy and social revolution was unanticipated and unplanned. It would have been invisible to forecasters taking snapshots of the U.S. economy as the twentieth century dawned. At the start of this century, with technologies in hand, growing markets, and blueprints on how to transform U.S. energy use, it is reasonable to assume that renewable energy and steep reductions in CO_2 emissions can be with us before my daughters reach my age. But to do that, we need to get beyond simple snapshots and expert projections that do not account for rapid social changes or the role of citizens like you and me. Few people would have predicted when I was in high school or college—when smoking was still stylish and doctors and athletes endorsed tobacco—that today only a small minority of Americans would be found out in the cold puffing in front of doorways.

We have already seen the damage that fossil fuels are doing to individuals and communities in the United States, let alone the rest of the world. I have argued, too, that we have the organizations, resources, and momentum to demand further change. That's why, still using the public health approach, I want to look at signs that a very different future, based on renewable energy and efficiencies, is not only possible, but already emerging. These developments hold out hope. But none is inevitable. They depend on changes in the political climate and on our own actions. When we fast-forward to 2030 or 2050, no one can be certain what our energy needs and sources will be. But we ought to jettison the prevailing, conventional assumption that our future will still be dominated by fossil fuel. There are too many indicators that we can build a safe, secure, and renewable energy future.

Renewables: Wind and Solar Power

To make sure that the United States switches to more efficient, renewable, carbon-free energy sources, we will need to move boldly. This time, unlike the developments that gave us gasoline, the auto, and highways, we can plan for it. Sweden, for example, has already announced that it will move to a carbon-free society by 2020.[3] The old, noisy, polluting, smoke- and CO_2-belching industrial engines of the nineteenth and twentieth centuries will be gone. Sure, Sweden is a small, homogeneous society filled with tall, blond, dour folks with a population half that of Michigan. But they do drive Volvos and Saabs. They have highways, produce steel, make movies, use computers, and eat breakfast over there. What distinguishes Sweden and other European nations that have taken the lead away from the United States in producing wind turbines, green buildings, mass transit, and more is a combination of technology, investment, and, especially, government policy. These reinforce and drive consumer and business practices toward efficiency and renewables.

Three things are needed for rapid change—technology, investment, and public policy. They are mutually reinforcing. Where there has been public policy promoting investment, technology has followed. When more investment in technology has been made, public policy has come right behind. A graph Chris Flavin shows is simple. Dotted lines with an arrow moving in an unending circle: investment → technology → public policy → investment. Why is tiny Denmark a world leader in wind turbines? Because its government created policies that promoted investment in wind power. As investment has grown, new, more efficient turbines are being produced. More investment and new policies follow.

Each of the major renewable technologies is poised to take off. Wind power, generated by modern, increasingly efficient turbines with long graceful arms made out of high-tech materials, has increased globally by 12 percent in a decade. The size of such power generators has already increased from a typical 50 kilowatts (kW) in 1985 to as much as 3,000 kW in 2005. In 2004 alone, 8,210 megawatts (MW) of wind energy capacity was added around the globe, bringing the total to 47,760 MW or enough to power 22 million typical European homes.[4]

Europe is still the leader in global wind power, with 72 percent of capacity. Germany and Spain account for more than half of world capacity. But the United States ranks third, and with the extension of a federal tax credit in 2005 and 2006, record growth of more than 2,400 MW of new installed wind capacity occurred in both years. This compares to slightly less than 10,000 MW of new natural gas plants brought on line in 2006 and less than 400 MW of new coal- and oil-fired generating plants. In sum, for more than two years, wind power has ranked second to natural gas and is growing rapidly. Major corporations like General Electric are making significant investments.[5] Critical to continued growth will be extension and increases in federal production tax credits for wind. Even now, the cost of wind energy, at 4–7 cents per kilowatt hour (kWh), has sunk below the cost of natural gas. This was even before the turmoil in Iraq and the Middle East sent gas costs skyward from 6–7.5 cents per kWh in 2006. And wind power is already as cheap as coal, whose costs range from 4.5–5 cents per kWh, even without factoring in health costs and environmental damage associated with coal mining and combustion.[6]

Solar power, which had a burst of interest and investment during the oil crises and high prices of the 1970s, failed to sustain its promise, especially as oil prices subsided and federal policies failed to subsidize or promote solar. Since that time, however, solar technology has been vastly improved (solar panels are made up of photovoltaic cells, or PVs for short). Rooftop electricity has come down in price from $30 per watt in 1975 to $3. The result is that solar PV capacity on American rooftops has more than tripled since 2000. Solar panels are also readily available, as home improvement giants like Home Depot now stock and promote them routinely. Looked at over time, U.S. solar PV–generating capacity is booming in terms of growth, if not yet absolute share. Solar power has grown from practically zero gigawatts (GW) in 1980 to about 1.75 GW in 2000 to about 4.3 GW in 2004.[7] The upward curve is steep.

There are other promising energy sources on the horizon that can combat climate change: tidal and wave energy from the ocean and rivers, geothermal heat from beneath the earth, and more. These are already being deployed in parts of the United States, Australia, Indonesia, Iceland, and beyond.[8] Early indications are showing real success, affordability, and the potential for rapid growth and investment.

There are a few potential drawbacks, however, to wind, solar, and other renewables that provide electricity. Wind farms have already been opposed in a number of cases by local residents and some environmentalists. Objections range from bird and bat destruction, to potential noise from turbines, to aesthetic questions over scenic views and sailing areas. The most serious impact on birds from wind blades has occurred primarily at a single, four thousand–windmill site covering 50 square miles in the Diablo Mountains near Altamont, California. More than 4,700 birds have been killed, including federally protected golden eagles. After lawsuits were filed, some operators, like G3 Energy, have already begun to replace huge, early-model turbines with smaller, less dangerous ones. Ultimately, portions of the facility will need to be entirely replaced and moved.[9] Although bird and bat kills have generated opposition elsewhere, new wind farms now generally carry out studies of local birds, bats, and migration patterns and work to minimize damage. Supporters of wind power, meanwhile, point out that millions of America's birds are destroyed by overdevelopment, cars and trucks (60 million–80 million birds), household pets, tall buildings, home windows and towers (between 100 million and 1 billion birds), and agricultural pesticides (67 million birds).[10] Most leading environmental organizations support the expansion of wind power, but with careful reviews. Typical is National Audubon Society President John Flicker. He says: "Audubon strongly supports wind power as a clean energy source that reduces the threat of global warming. Location, however, is important. . . . Each project has a unique set of circumstances and should be evaluated on its own merits."[11] The most highly publicized single case has been the Cape Wind offshore Massachusetts wind farm in Nantucket Sound between Cape Cod and Martha's Vineyard. It has been delayed by significant political battles with passionate forces arrayed on both sides, including opposition from the Kennedy family. Some environmental groups like Massachusetts Audubon, tentatively favor the project, but only after further study and mitigation plans for wildlife are put in place.[12] Meanwhile, major offshore wind farms have moved ahead in Texas and elsewhere. In Delaware, where offshore proposals have drawn wide support, marine scientists at the University of Delaware have suggested a number of ways to reduce harm to marine mammals and birds.[13]

Another concern about wind and solar energy has been intermittency, the fact that the sun does not shine all day, nor does the wind always blow. The U.S. electric industry currently relies on large, central power stations, including coal, natural gas, nuclear, and hydropower plants. Together they generate more than 95 percent of the nation's electricity. Night and day, they give you and me a steady source of power. Most electric utilities operate a combination of so-called baseload plants (often coal and nuclear) that operate most times while others (often natural gas) are utilized only when demand is high. Given their age, health and security risks, and power outages, these twentieth-century sources need to have backups and then be replaced. Over the next few decades, renewable energy could help fill that need. Some renewable power plants can provide steady power whenever it is required—using geothermal, concentrating solar (with storage), and bioenergy sources. Yet even intermittent sources add significant value by providing electricity when it is most needed and most costly to produce. In many parts of the country, periods of peak sunlight coincide with peak power demand for air-conditioning. Wind resources can already be forecast at least two days in advance, and fluctuations in power can be reduced, if not eliminated, by spreading solar or wind generators across a sufficiently wide region. In the future, new technologies like advanced gas turbines and fuel cells, as well as new storage devices, will permit much higher percentages of intermittent generators.[14]

Interestingly enough, there is a storage device that can hold excess power produced by renewables during sunny and windy periods, resolve doubts about intermittency, and help rid the United States of oil dependence. It's called a car battery. New high-tech ones are being planned and produced to turn the current generation of fuel-efficient hybrid cars into plug-in hybrids. When powered by electricity from renewable sources, these create no emissions. Let's take a look, then, at cars and what fuels them.

More Energy Efficient Cars and Fuel

Hybrids, using both gasoline and battery-powered engines, as suggested by *Ending the Energy Stalemate* and other policy reports, have been the first, sensible step. They have gone, while I was conceiving of and writing this book, from prototypes to popular products like the Toyota Prius. When the Prius came on the market in 2000, it sold fifteen thousand cars. By the end of 2004, it was selling that many a month, with long waiting lists for more. Other companies soon followed, with Hondas and even Ford sport utility vehicles (SUVs). The Prius gets about 50 miles per gallon in an increasingly, roomy, stylish, and zippy car. That compares to the current average fuel economy of the whole U.S. fleet of 25 mpg, a figure that has shown little progress from the 1970s.[15] As a result, in 2006, Toyota passed the Ford Motor Company as the second-leading car company in the United States. When gas prices spiked up as a result of the Iraq War, Americans starting buying and demanding even more hybrids. But government policies actually restricted the energy tax credits for which U.S. taxpayers could qualify by buying one of these cars to only the first sixty thousand vehicles sold and delivered by each auto manufacturer during a limited period of time in 2006 and 2007.[16] It is not enough that celebrities like Leonardo DiCaprio and Brad Pitt each own and drive two Priuses and tout their virtues. A change in policy is needed to further stimulate hybrid vehicles in the United States—a technology that will soon be further improved.

Gains against global warming through automotive and fuel changes have and will continue to come in stages. We will need to further increase mandatory fuel efficiency (CAFE standards) beyond the 35 mpg (but not until 2030!) to which the Congress in 2007 was finally able to raise them—over the opposition of the auto industry and a threatened Bush veto. In the meantime, we will need to keep moving from gasoline to biofuels like ethanol made from corn and beets and other plants. Perhaps later we will move to hydrogen as a fuel, though it presents significant new problems. Then we must complete the switch from our current internal combustion gasoline engines that spew CO_2 and air pollutants (and leak oil as well) to hybrid cars. Hybrids come in many different kinds, but the most efficient work serially, moving from a gas to an electric engine while gathering energy from braking and storing it in batteries. The next phase will be to move to plug-in hybrids that have larger electric engines and new, larger lithium ion batteries for storage. Ultimately, we should end up with all-electric cars powered by renewable electricity—not made by coal plants or nuclear facilities.[17] Finally, we can move to "hypercars." These are designed with far different technologies, materials, and styling that reduce weight and drag and enhance other performance qualities. As pressure has begun to mount around clean air, climate change, and energy independence for the United States, each of these improvements has begun to advance. But in each case, pressure, first from environmental groups and citizens, then from government, has been essential.

Even the threat of future mandatory restrictions on carbon in the Clinton years helped push manufacturers toward lower emissions and hybrids. I attended a number of ceremonies in the small auditorium of the ornate Old Executive Office Building, when Al Gore or Clinton himself first touted and then announced the NLEV (New Low Emissions Vehicles) program. The program was voluntary, but the White House, backed by environmental groups, clearly had tougher measures up its sleeve. This was fairly obvious to the largest collection of beefy auto guys that I had ever seen. Behind it all, was the idea that General Motors (GM) and other U.S. manufacturers had better get with the NLEV program now. Actual mandates are not far behind. Further stimulus was given by places like California, one of the largest car markets in the world, where low-emissions and then *zero*-emissions standards were passed. The auto companies and the oil lobby fought these so-called Pavley standards (named after their sponsor, State Representative Fran Pavley), which were among the critical and underappreciated measures fought for by national environmental groups. The Bush administration ultimately went to court to get such stricter state standards blocked. The Supreme Court finally ruled against the administration in late 2007, holding that the EPA had to regulate emissions. But EPA Administrator Steve Johnson then moved again to block implementation of California's stricter rule. Nevertheless, such efforts in California—accelerated by the power blackouts of 2003 and continuing human health costs from smog—have over time yielded tremendous results. Although California's population and number of car miles driven has increased dramatically since the 1990s, emissions, thanks to efficiency measures, have remained steady.[18]

Such pressures moved even the most staid of auto manufacturers, GM, and its ally on climate and air issues, the United Automobile Workers (UAW). Both have fought clean air and CAFE standards and CO_2 caps of any kind. With its industry standing and jobs at stake, GM poured $1.2 billion into hydrogen fuel cell cars, with the first one hitting the showrooms in 2006. This is a massive bet, because a number of obstacles remain before you and I drive to the bookstore (actually your doctor says to walk, jog, or cycle there) in a glistening, hydrogen-powered Cadillac.[19]

A key issue for global climate change and government policy is where does the hydrogen come from? Hydrogen is created by the electrolysis of water. An electric current separates out the H in the familiar H_2O. But most electricity, as we have seen, comes either from dirty, old, coal-fired utilities, from cleaner, but CO_2-emitting natural gas ones, or from nuclear power plants filled with radioactive uranium and wastes. Even as the current Bush administration fights mileage standards, its energy proposals call for increased attention to hydrogen fuels. This assumes that hydrogen will come from larger numbers of nuclear power and fossil fuel plants. If so, when someday you buy your proud, no emissions, hydrogen-powered Cadillac, it would be backed up by a long, seamy trail of excessive energy use, carbon emissions, and radioactive waste. Any real hydrogen future will need to ramp up sources of renewable electricity. Hydrogen can be made quite transportable in liquid or solid form, even running through old oil pipelines. But it is potentially explosive. Even if such fuel can be made relatively safe and secure and produced from renewable power, there is, at present, precious little infrastructure—hydrogen storage and filling stations—at which to tank up. Such changes will take even more investments, more technology and *policies* that encourage and goad auto companies, especially ailing ones like GM and Ford.

In the meantime, much of the interest in alternative, renewable fuels for cars (and other vehicles) centers on what are called biofuels. They come in two kinds. The most common is ethanol. It's produced from sugar or starch crops such as corn. The other is biodiesel, which can be produced from either vegetable oils or animal fats. Interest in biofuels has increased as oil prices have soared, and some initial environmental problems appear to be improving. A major concern had been whether or not, in net terms throughout the entire production process, more energy would be used than actually produced from ethanol. Such a result, stemming from counting fossil fuel–based products like fertilizer, tractor fuel, and processing energy, would actually increase overall greenhouse gas emissions and *worsen* global climate change. Nevertheless, recent advances in converting crops to ethanol and the increased use of biologically produced energy have now resulted in a net energy *gain* and a reduction in greenhouse gas with ethanol. But the near future, according to many analysts, is even brighter with the advent of cellulosic ethanol. The ethanol is produced from cellulosic crops like switchgrass, from agricultural wastes such as plant stalks and leaves, or even from woodchips; alternatively, these same raw materials can be transformed into synthetic diesel produced through a method pioneered in Germany and South Africa. Here biomass is first gasified into carbon monoxide and hydrogen, and then, through the chemical synthesis of the Fischer-Tropsch technique, it is transformed in liquid fuel such as diesel. Though expensive right now, these methods are about to be introduced on a commercial scale.[20] If cultivated properly, cellulosic crops can actually sequester carbon as well, further helping to reduce the CO_2 in the atmosphere. Nevertheless, as the Worldwatch Institute warns, if biofuels are to be successful, world production must avoid low-yielding crops and the use of fossil fuel–produced energy. Real care must be taken not to drive up food prices or destroy ecologically sensitive areas.[21]

With these caveats in mind, one can imagine a very short transition period to biofuel-based cars, especially as new models are designed to accept gasoline, ethanol, or blends. World production of ethanol has already more than doubled between 2000 and 2005, while production of biodiesel quadrupled. Again, it is the trends that count. Biofuels are currently about 1 percent of transportation fuel, but production of ethanol rose 19 percent in 2005 and biodiesel an amazing 60 percent. As oil prices have risen, investment in biofuels has accelerated in Brazil, Europe, and the United States, with major commitments to biofuels in China, India, Columbia, the Philippines, and Thailand. And Sweden, as we have seen, has stated that it will end its dependence on fossil fuels by 2025 with biofuels playing a major role.[22]

The Brazilian picture is important because, following the formula of public policy to investment to technology, Brazil has become the world's largest producer and user of ethanol. That has led to a 60–80 percent reduction in its production of greenhouse gases. This feat has been accomplished in Brazil by a remarkable shift in sales of flexible fuel vehicles, or FFVs, from 4 percent to 80 percent of all auto purchases within three years. FFVs are relatively simple to produce with today's engines (and further advances will increase efficiencies) and can use gasoline or ethanol. The result in Brazil has been a reduction of petroleum use for cars and light trucks of 40 percent and a savings of $50 billion in oil imports. In the United States, there are already more than 6 million flexible fuel vehicles and a supply of 4 billion gallons of ethanol being blended into gasoline. And in California there are almost as many FFVs as diesel vehicles. Production costs? In the United States ethanol production is currently $1.00 per gallon versus anywhere from $1.60 to $2.20 a gallon for gasoline. The curve of production? It already shows a steep 20 percent increase.[23]

Once cellulosic ethanol becomes more common and processes for biomass such as using grasses like *Miscanthus* that can produce more than 20 tons of ethanol per acre are widely used, U.S. ethanol use can rise sharply. Entrepreneur and biofuels advocate Vinod Khosla describes it as "turning South Dakota into the Saudi Arabia of ethanol."[24] Forty-four million farm acres are already producing 857 barrels of ethanol a day in South Dakota; improved methods could produce 3,429 barrels a day. That would put South Dakota in third place behind Saudi Arabia, closing in on Iran's second place. The Ceres Corporation has estimated that if the United States turned 78 million of its agricultural crop acres currently used for export and 39 million acres of unused crop land into biomass fuel crops, by 2015 we could be producing 384 million gallons of ethanol a day, or 75 percent of current U.S. gasoline demand.[25]

But, my realist and environmentalist friends are muttering, cellulosic ethanol depends on breakthroughs in genetically modified enzymes to break down that cellulose. They're not ready yet. And though pollution is much reduced and greenhouse gas emissions brought down to a minimal level, the result is *not* zero. And if we keep expanding the auto economy worldwide, biofuels risk competing with actual food for the finite areas of agricultural land that the world (and the United States) has. Fair enough. Corn, beets, switchgrass, and wood are all renewable, but there are limits to how much can be grown when measured against the need for food. And more research is needed for further breakthroughs. This argues for sensible policies, both in the United States and internationally, to ensure that biofuels do not start to harm the already very hungry parts of the world. In fact, not long after President Bush committed the nation to substituting 20 percent of the petroleum we use with ethanol and visited Brazil to stimulate a trade deal and an "OPEC for ethanol" doubts began to surface. As investors began to move swiftly toward ethanol in ways not predicted by conventional analysts, potential environmental problems became more real. One of the original sponsors of Brazilian legislation that stimulated ethanol production, Fábio Feldman, citing cane plantations "the size of European states," has already called for slowing down and reviewing the impacts from the boom.[26] Environmental groups have also begun to report that crop production for ethanol is beginning to eat up new, large sections of wooded savannah—home to jaguars, blue macaws, and giant armadillos—in Brazil's Cerrado plateau.[27] In sum, ethanol holds promise. It can reduce our need for oil and reduce greenhouse emissions. But it has its limits. Its growth will need restrictions and regulations. Ethanol is not a "silver bullet."

The electric car is perhaps the best possibility. My mom told me many times of the fancy electric car driven by a wealthy neighbor in Philadelphia in the early 1920s. There were plenty of trolleys; some horse-drawn vehicles were still in use; and gasoline cars, mostly Fords, were not yet ubiquitous. Fifty years later, each morning outside my window in Philadelphia (not far from a commuter train from the 1870s and trolleys on Germantown Avenue), my neighbor unplugged his battery-powered Volkswagen Rabbit, on loan as a PR gimmick from the Philadelphia Electric Power Company, and sped off down the street. Whatever happened to the electric car? One organization on whose board I served, the Environmental Alliance, held fundraisers during 2006 to back political candidates who supported energy efficiency. The events featured the documentary *Who Killed the Electric Car?* The answer is the same combination of short-sighted corporate moguls in the gas and auto world who destroyed trolleys, gave us huge cars with fins, and then suburban SUVs.[28] But increasing oil prices, both regulatory and popular pressures for more fuel efficiency, and concerns about oil security, global climate change, and the staggering health costs from pollution have already begun to change that.

Plug-in hybrids allow for a smoother transition to the all-electric car and are developing support from across the political spectrum. They improve on a normal hybrid by adding more batteries, a charger, and an electric plug. They use the Toyota-style series parallel drive train with the extra batteries allowing for greater time away from the gas engine and hence more efficiency and fewer emissions. No new infrastructure or service stations are needed, as you can plug into a normal 110-volt socket in your home or garage. With just an overnight charge, a plug-in can average better than 100 miles per gallon. If not recharged, it still runs as a regular hybrid, getting 50 mpg. In 2007, Toyota announced that its first plug-ins would arrive with the 2008 model, though the date has now slipped to 2010.[29] But the race is on. At the popular 2007 Detroit auto show, General Motors unveiled its prototype Chevrolet Volt, an electric car that, when in full production, can save consumers about $900 a year on gas, be recharged overnight, and cut 4.4 metric tons of CO_2 emissions annually for the average driver.[30]

There will be other options for cars in the future, too. Amory Lovins, whom my producer, Beth Parke, had to cajole me into airing on my radio program, *Consider the Alternatives,* has been gaining acceptance by the mainstream. By 2004, Lovins and his coauthors had actually received Pentagon funding toward their study of energy efficiency published as *Winning the Oil Endgame: Innovation for Profit, Jobs, and Security.*[31] It details how to move to "hypercars," which use advanced, space-age materials, hydrogen power, and better designs to drastically reduce weight while maintaining strength and increasing aerodynamics and fuel efficiency.

The Built Environment and Sprawl

Okay. We can get an end to coal-eating, pollution-belching power plants through renewable energy sources. Increasingly, we can have efficient hybrid cars using low-emission fuels as we work toward zero-emission electric vehicles. And so far, investors make money, U.S. technology is enhanced in world trade, jobs are created, and CO_2 emissions and unhealthy pollution reduced. It is beyond my scope in this chapter to deal with other transit, but suffice to say that we could vastly improve our rail systems. China has already invested billions in maglev (magnetic levitation) trains, which float above the ground on supermagnets and whiz along at speeds more than double those of regular trains.[32] And there is growing interest in light rail. All of the light transit lines in the United States in major cities, with a few exceptions, disappeared within less than ten years. Surely we can get them back as quickly. For younger readers, these were "trolleys"—short, trainlike vehicles with a rectangular roof, running on tracks, powered from overhead electric lines, and immortalized by a song with the lyric, "Clang, clang, clang!" A few examples can still be seen in parts of older cities like Philadelphia and Baltimore, and newer, hipper trolleys are appearing in younger, "green" cities like Portland, Oregon. But the General Motors Corporation, in league with oil companies, eager to sell its diesel-powered city buses, bought out and eliminated the trolleys and the song.[33] We were left with increased urban air pollution, noise, stench, and rising rates of disease, especially near bus terminals in places like Harlem.[34]

With cheap energy in the postwar period, increasing affluence, and the availability of land, the United States saw significant growth in its cities and suburbs, and now in its rural and semirural areas. The phenomenon is summed up in the properly ungainly term "sprawl." Because I grew up on Long Island after World War II when my parents moved from Manhattan for the schools and the shady suburbs, I know sprawl when I see it. I remember Kennedy Airport when it was Idlewild and mostly wetlands and Quonset huts. I watched truck farms in Nassau County disappear as Levittown was built. Nassau sported signs that proudly boasted we were the fastest-growing county in the country. Then came the Long Island Expressway that was to relieve traffic congestion, the mall that replaced the fields I played in from which Charles Lindberg took off on his historic solo flight to Europe, and, finally, air and water pollution that caused my mom to say when I started at PSR more than fifteen years ago: "You're going to deal with breast cancer, aren't you?" This phenomenon of unchecked, unplanned growth has spread (sprawled?) across the country. It has reached formerly pristine areas like Arizona, New Mexico, and other parts of the Southwest and western mountains that in my youth had very low populations. Retirees, old Air Force generals, and asthma victims once went to Arizona for the open spaces, pure air, and sunshine. Now, like Texas, places like Phoenix and Tucson are increasingly polluted. They suffer from highway congestion and rampant sprawl. In Arizona, where I have birded for vermillion flycatchers along the San Juan River estuary, hidden in writers' retreats near the Saguaro National Cactus Forest, and been dazzled by diving hummingbirds in the Madeira Canyon, battles between developers and environmentalists are growing. If we are to combat global climate change (which will surely destroy American culture in the Southwest as it did the Pueblo and Anastazi people before us), we need hopeful signs that growth and the design of spread-out cities, suburbs, and exurbs can be controlled and contained. Because sprawl creates a variety of health problems, economic disparities, and other social justice issues, it has given rise to movements concerned with environmental health, safety, racial justice, and the quality of life that could add political clout to calls to prevent global warming.

Numerous national environmental groups and coalitions focus on the nation's transportation policy, especially the massive highway bill that propagates sprawl. In late 2005, Congress passed the $286 billion act with the impossible acronym SAFETEA-LU. It will need to be renewed in 2009. Groups like the Surface Transportation Policy Project (STPP), Smart Growth America, Scenic America, and the Rails-to-Trails Conservancy, with help from other Green Group members and health groups like PSR and the American Lung Association, fought to reduce support for paving and car space found in the highway bill at a cost in the billions. These mainstream national coalitions and groups also have grassroots components and enthusiastic members who can be found nationwide. They fight to increase the funding and policy measures needed to promote green space, bike paths, and alternative transit that make up a small portion of this huge federal piece of legislation that has been backed since the Eisenhower days by powerful interests in construction and autos. The huge network of interstate highways we enjoy was passed under the guise of national defense at the height of the Cold War so that massive federal subsidies would be accepted during a "conservative" Republican administration. The American Road Builders' Association had begun lobbying in 1943 for a national interstate highway system even before World War II ended. As my colleagues Howie Frumkin and Dick Jackson described it in *Urban Sprawl and Public Health:* "Their coalition included oil, rubber, asphalt, and construction industries; car dealers and renters; trucking and bus concerns; banks; and labor unions." When the Interstate Highway Act was finally passed in 1956 it was funded from an inflexible source. The system was paid for by pouring gasoline taxes into nondivertible highway revenues, leaving the United States with "the world's best road system and very nearly its worst public transit offerings."[35] If we count the costs of air pollution and some forty thousand to sixty thousand deaths per year, plus injuries; the destruction and neglect of trains and mass transit that followed; and the loss of small towns and local economies across America, our highway and car culture has cost us immensely. And it continues to promote exurban sprawl, eating up agricultural and scenic lands. Like you, I am not quite ready to give up the convenience and freedom of a car. But enough is enough. A national commitment on a Cold War scale is needed to transform American transportation and the communities fundamentally affected by it. Imagine not only zero-emission hypercars on our highways but also maglev trains, linked to urban commuter and light rails; efficient, hourly Zipcar rentals; trolleys; and denser, renewed cities and suburbs that could be walked and biked in. You begin to get just a small picture of the positive energy future in America that we can have in our lifetimes.

Reducing sprawl is essential, first of all, to reducing commuting to urban work centers by car and endless short trips in the suburbs by cars, vans, and SUVs—to school, to shop, to visit friends, to do almost anything. Each gallon of gasoline burned in your car produces about 20 pounds of CO_2. When greenhouse gas emissions were measured in the United States in 1990, industry was the largest source of carbon emissions. But because industrial emissions have declined while transportation emissions have increased by about 2 percent per year, by the year 2000 transportation (mostly cars and trucks) had become the largest source of greenhouse gas. And it is in the category of light trucks, the definition given to SUVs through a compromise for the Clean Air Act that spared them tighter fuel-efficiency standards, that CO_2 emissions have increased the most—more than 50 percent between 1990 and 2001.[36] As gasoline prices have increased, Americans have begun to move away from gigantic Ford Explorers and other behemoths. But stemming the upward trend in the numbers of short car trips and commutes to work will be necessary, too. In the Washington area where I live, daily commuting time and sprawl, traffic congestion, and miles commuted are among the worst in the nation. Developers and highway lobbies still have the upper hand, as witnessed by the widening of the Wilson Bridge across the Potomac River to twelve lanes and the approval, after years of opposition by environmentalists, neighborhood groups, and antisprawl advocates of an interstate highway connector across suburban Montgomery County to "ease" commuting to the eastern side of the county and on to Baltimore and points beyond. Nevertheless, Washington, D.C., itself has been undergoing a renaissance, with increased urban population for the first time in decades. There are significant new housing complexes near an expanded Metro system, or within bus or walking distance of work and weekend whoopee. Light rail is being built, and bike trails have increased along with bicycle commuting. Significantly, the 2006 election put a number of the area's smart growth advocates in power, including county council member Ike Leggett as Montgomery County executive. Leggett, a genial, long-serving, environmentally friendly African American, beat the early favorite, who had been backed by developers with more than $1 million in contributions.[37] Other signs are emerging, too, including a final decision to build a rail system connected to the Washington Metro out to Dulles International Airport, one of the nation's largest national and international hubs. At present it can only be reached by car or cab.

Such changes away from car-centered, sprawling cities and suburbs will be increasingly essential, as the health effects of air pollution from autos and coal-fired power plants remain a major U.S. problem. As we have seen, they give off not only carbon emissions causing global climate change; they also produce ozone, tiny particulates, and other pollutants that kill as many as forty thousand to sixty thousand Americans per year. But even inside their own vehicles, drivers and passengers are at risk, as are people who live in more sprawling neighborhoods. One epidemiological health study in Seattle compared various neighborhoods based on urban form, density, connectivity, and land use mix. It found that people who lived in conditions typical of sprawl were exposed to greater air pollution emissions such as NOx, CO, and volatile organic compounds (VOCs). Studies in the United States and Asia also found that VOCs like benzene, styrene, toluene, and formaldehyde—which evaporate from gasoline, are carcinogenic, and are found in exhaust fumes from nearby vehicles—were higher in heavy traffic, older cars, and those with warm interiors from sitting in the sun or using heaters. In some circumstances, VOCs inside vehicles actually *exceed* those of outside air pollution. Frumkin and Jackson report on a particularly chilling 2001 report by the Natural Resources Defense Council and the Coalition for Clean Air. Diesel exhaust in California school buses reached up to four times higher than in cars traveling nearby. The study estimated that based on the observed exposures, the children's lifetime risk of cancer was increased twenty-three to forty-six times above significant levels as defined by the EPA. Similar results were found in Connecticut and in another 2003 California study.[38]

Battling sprawl is increasingly a concern to minority communities. Numerous studies have indicated they suffer disproportionately from air pollution and other environmental ills. Robert Bullard, a sociologist from Clark University in Atlanta, has written widely on the subject and is at the center of movements for environmental justice. As he and his coauthors put it in *Sprawl City: Race, Politics, and Planning in Atlanta:* "Most of the reports on the topic gloss over or minimize the social equity implications of Atlanta's sprawl problem, which is intricately linked to both race and class. Atlanta's history is steeped in racial politics; every public policy decision made in the region operated under this backdrop."[39] The same can be said for most sprawling American cities. The automobile, de jure and de facto segregation, and the rise of suburban economies have created unequal exposures to good schools, jobs, safe streets, and clean air. Nevertheless, antisprawl coalitions continue to form increasingly across race and class lines—in cities, suburbs, and nationally. And many trend-setting cities have already provided examples of controlled green growth. Portland, Oregon, for example, has created a green zone around the city that restricts development while increasing its emphasis on urban transportation, housing, and green buildings.

Green Buildings

To reach a carbon-free America, we also need to see further reductions in CO_2 emissions through the redesign, retrofitting, and creation of entirely new, energy efficient buildings. We have a historic opportunity to revolutionize the marvelous infrastructure of homes, offices, and factories throughout the United States. Most of our buildings and factories currently use large amounts of fossil fuel energy (and waste the majority) for heating, cooling, lighting, and machinery. All told, this "built environment" uses a significant portion of all U.S. energy and emits large amounts of greenhouse gases. We have already seen how enlightened businesses like Toyota have moved to increase the efficiency of their production processes. But to imagine an entire landscape of change requires real vision, a revolution if you will.

Bill McDonough will do for a contemporary American visionary and revolutionary, even if he is an architect with conservative bowties, blazers, and a bit of Bill Buckley's mandarin style. McDonough was the dean of the University of Virginia School of Architecture until devoting himself full time to his own architecture firm, William McDonough and Partners, and his commercial product and process design firm, MBDC.[40] I first met McDonough during the 1990s at a Washington political fund-raiser sponsored by a newspaper heiress in a rather impressive traditional mansion overlooking the Potomac in leafy McLean, Virginia. Before long, he and I were happily talking about architecture, and incredibly enough, McDonough's sermon offered at the Cathedral of St. John the Divine in New York City. McDonough had preached at its annual blessing of the animals, a now traditional environmental service. His combination of conservative appearance, missionary zeal, and hopeful vision made a deep impression. I had not heard of green architecture at that point. Nor did I realize that McDonough would prove a rather effective prophet of proper design, profitability, and planetary salvation. His designs have ranged from a refurbished River Rouge Ford assembly plant in Dearborn, Michigan, to residential homes, to the new School of International Service building at American University where I teach.

I have since appeared alongside McDonough and watched him mesmerize and convert even the most difficult of audiences. They give him standing ovations as he waxes eloquent about design principles based on nature; everything is used and reused and, like a tree, relies on all its energy from natural sources like sunlight. McDonough is given to metaphor. My favorite can be seen in a documentary film that features his ideas, a sort of cult classic for those who have been worried about global warming from early on. In it, he describes the ocean crossing of nineteenth-century Yankee clipper ships with sails to the wind. They used no fossil fuels, were built of renewable timber, and were examples of elegant, efficient design. Only a few years later, they were replaced by iron steamships. These belched smoke, consumed resources, and forced crews to work in unspeakable conditions shoveling coal below decks. And then comes the clincher. With its poor design and excessive use of energy, our society is a steamship; we all work in overheated, poor environments just to keep it going.[41]

McDonough's ideas have now begun to spread—and not merely to pioneering, modest edifices like the environmental sciences building at Oberlin College touted by another green building advocate, David Orr.[42] Manhattan, the ultimate in skyscrapers and energy-intensive, modernist, sealed, glass twentieth-century buildings, is now the proud home of two major green skyscrapers for Condé Nast and the Bank of America, with more to follow.[43]

But the trend toward green buildings is not limited to university buildings or skyscrapers. In less than a decade, green building has moved from cult figures and prophets like McDonough to the front pages of *USA Today.* Reporter John Ritter could write in the middle of 2006 that "Building 'Green' Reaches a New Level." Always on the environmental cutting edge and a trendsetter, Portland, Oregon, is now selling green condominiums overlooking the Willamette River with views of Mt. Hood and Mt. St. Helens. They are part of a $2.2 billion South Waterfront project that is going up on an old industrial site linked to the city center by a new streetcar line. To officially qualify as green, new public buildings must meet LEED (Leadership in Energy and Environmental Design) standards that are set by the U.S. Green Building Council and required by the federal government, fifteen states, and forty-six cities. Four states and seventeen cities now offer incentives for LEED-rated private buildings, and Chicago and Pasadena, California, now fast-track permits for green builders.[44]

Green buildings, which were also pioneered and promoted by environmental groups like NRDC, Environmental Defense, and the National Wildlife Federation that have green headquarters, must qualify as green in their building materials and their energy and water use. The condos in Portland sport nontoxic paint and finishes, wheatboard cabinetry, Caribbean walnut and other plantation-grown wood, features such as rainwater capture for low-flow toilets and landscaping, and, of course, high-efficiency heating and cooling systems with panels that serve as both sunshades and solar power generators.

Other multibillion-dollar green projects are in progress, including the redevelopment of the Camden, New Jersey, waterfront along the Delaware River across from Philadelphia. It has been a decayed wasteland since American companies like Campbell Soup moved out decades ago. Camden still maintains a small, lonely, historic house in the area where poet Walt Whitman once lived. Whitman, who sang of uniquely American places, walked everywhere, and overlooked Manhattan from his perch at the *Brooklyn Eagle* where he was a reporter, would surely approve of these new democratic vistas. New York City's Meadowlands development area is also going green, but the trend is not confined to pricey neighborhoods or condos like South Waterfront where, Ritter reports, Edward and Michelle Walsh paid $790,000 for a two-bedroom, three-bath unit with den after selling their Arlington, Virginia, home and moving to Portland.[45]

Seattle's equally fashionable High Point neighborhood now has the nation's first green public-housing project as well as six hundred apartments and townhouses, surrounded by green traditional houses, all selling at market rates. And, again, it is a mainstream environmental organization, the Natural Resources Defense Council, which has initiated a nationwide partnership with the Enterprise Community Partners. Called the Green Communities Initiative, it will develop low-cost, green housing, especially in urban areas. Begun in late 2004, the project is a five-year, $550 million commitment to build more than 8,500 environmentally friendly, affordable homes across the country. Enterprise Community Partners has already created attractive, affordable homes and apartments like the Wellstone in Minneapolis. Such green apartments and homes are located conveniently to urban transportation, in walkable neighborhoods, and they slash energy use (and greenhouse emissions) by at least 30 percent. Their location in denser neighborhoods also reduces sprawl and can cut emissions and pollution from commuting by car up to 50 percent. Such green housing also reduces water usage by 50 percent, saves forests by using recycled wood, and, by using nontoxic materials, reduces exposures and ailments related to traditional glues, sealants, and synthetics. And lower initial costs, reduced utility bills, and cheaper commuting costs make such initiatives ideal for lower-income families.[46]

For developers, the cost of a green building is usually no more than 1–2 percent higher than that of buildings using traditional materials and technologies, with costs declining as demand increases. For businesses and commercial buildings, energy savings alone quickly make up for the initial added cost. Once again, green building is a small share of America's overall construction, but the trend line is moving rapidly upward. At least 6 percent of nonresidential construction, a $15 billion-slice of the industry, is already green, while a McGraw-Hill Construction survey in March 2006 predicted that green building would reach a tipping point in 2007, with two-thirds of American builders offering green homes.[47]

The revolution brewing in green building reveals yet another hopeful sign. Buildings—unlike national parks, wilderness areas, or other traditional areas of environmental concern—involve everyone. They create overlapping constituencies and interest groups. Initially, environmental concerns about buildings focused on toxic materials, sick building syndrome, and the pollution that came when synthetic materials were incinerated, buried, or leached chemicals into water supplies. Thus, the environmental health movement—which includes those who suffer from various diseases related to pollution—has been pushing hard for reduced use and elimination of various toxic products and materials in homes, buildings, and schools. Trailblazers like Phil Landrigan of Mt. Sinai Medical Center and Herb Needleman of the University of Pittsburgh have been warning for decades of the dangers of lead, mercury, pesticides, solvents, and other substances in homes and in the commercial and industrial world.

One early manifestation of the green products and building revolution was the coalition Health Care without Harm (HCWH). Started by leaders in the environmental health movement like Charlotte Brody and Gary Cohen, as well as Physicians for Social Responsibility, and nurtured by Michael Lerner, the founder of Commonweal, a holistic cancer treatment and retreat center, HCWH first tackled the problem of medical waste incineration.[48] A 1994 EPA reassessment report had named medical incinerators as the second-leading source of dioxin, a highly toxic by-product of burning any product that contains chlorine or its compounds. HCWH attacked the use of polyvinylchloride (PVC) plastics and other products, then went international; recently it has begun to promote green hospital buildings, as have growing portions of the medical establishment. Thus, the South Waterfront green building complex in Portland is anchored by a new bioscience center at the Oregon Health and Science University. It will be the first U.S. building to use chilled beams instead of conventional air-conditioning.[49] Similar green hospital facilities have also been completed at Emory University, which has committed itself to an entirely green campus, and at increasing numbers of other medical facilities.[50] As such new concepts have emerged, they have become increasingly holistic, integrated, and interdisciplinary. Green medical centers first began to eliminate PVCs and other toxic components, then looked to energy use and building materials, and finally to a broader concept of green medical care. It draws on innovative research showing that patients actually heal faster and better when given more natural views and surroundings that are also people- and family-friendly.

American hospitals in particular have become places of modernist, utilitarian, and high-tech laden, sterile architecture. People prefer to be at home. If they can't be, they prefer comfortable, airy, green, and human-scale environments with social interaction. And so medicine is increasingly turning to older notions of prevention and health promotion, not just clinical intervention and cure. Health care is one of our nation's largest industries. A "green medicine" approach not only saves energy compared with heavily air-conditioned, harshly lit, machine-laden hospitals, it can also join together physicians, nurses, health-care workers, and their unions with university officials, patient advocates, and families of all sorts. They can push not just for CO_2 reductions, but for an entirely new, hopeful vision for American medical care and our quality of life. But such a vision of a caring, human-scale future is tied, finally, to public policy. To be fulfilled, the transformation we need requires detailed plans and policies that can survive in the political arena.

Future Energy Policy Scenarios

Ever since 2001, when the Bush administration unveiled its first energy plan—drafted mostly by lobbyists and supporters of the oil, coal, gas, and utilities lobbies—the U.S. Congress has been locked in battle over how to proceed. The original Bush plan called for increases in oil production and drilling, the building of large numbers of new coal-fired power plants, increased supplies of natural gas, along with some slight lip-service and R&D funding for a little bit of work on renewable energy. All of this was backed, as it has been throughout the post–World War II period, by massive subsidies, royalties, and tax breaks for oil, gas, coal, and nuclear power companies.[51] Because opponents, including environmentalists, fought much of this, with a special emphasis on preventing drilling for oil in the pristine Arctic National Wildlife Refuge (ANWR—pronounced AN-wahr), the result for years was mostly stalemate in new energy proposals. Meanwhile, as the need to reduce greenhouse gases has grown, and the dangers of relying on unstable foreign sources of energy have become clear thanks to the failed war in Iraq, a variety of alternative energy plans have been put forth. The energy bills of 2005 and 2007, though loaded with carbon and subsidies for it, showed some improvements. President Bush at least began to shift his rhetoric. But when finally passed, the Energy Act of 2007, threatened with a possible presidential veto and still insufficient environmental supporters in

Congress, had been stripped of some very important provisions like help for solar energy. It had reduced, but not eliminated, massive subsidies for the fossil fuel industry, maintained support for the nuclear industry, and moved forward with plans for domestic oil drilling. But there were signs that real change is in the air. Backed by environmental groups and a growing climate movement, small but significant victories were won. A bill speeding up the demise of the incandescent light bulb and mandating compact fluorescents in the future was passed, as was the first new CAFE standard in many years. Once again, as in the transition from the first Bush administration at the start of the 1990s, the stage was being set for lengthy electoral battles and a fierce struggle for change when a new administration takes over in 2009.[52]

Given such continuing struggles over energy, which will continue long after President Bush, it is important to look closely at the thinking and the sorts of recommendations that policymakers take seriously. Such proposals generally are produced by reliable, well-known, if cautious, policy think tanks or national entities. The various authors and former officeholders involved are likely to have some sway with any new administration. And unlike political candidates, they gain their power and reputation from policy expertise and prudence, not passion or politics. With the Bush administration still in power until January 2009, most policy proposals to date have shied away from bold or optimistic scenarios. Nor have they based projections on increasing trends in new renewable technologies and markets. None have assumed that the political and policy atmosphere or climate will have changed dramatically thanks to efforts from folks like you. Let's see what they have to say.

Typical is the report from the National Commission on Energy Policy called *Ending the Energy Stalemate: A Bipartisan Strategy to Meet America's Energy Challenges.*[53] It is the result of huge amounts of research and discussion, and it grows out of an honest desire to get a better energy policy for the United States. Its goal is to slow as much as feasible the rate of growth of greenhouse gas emissions, then to reduce them. But in its drive for balance, bipartisanship, and policy recommendations that might be passed by a Congress in the near future, the commission sets its sights too low. The burden, I believe, should always be on well-meaning realists to show how and why their cautious, moderate proposals would generate the popular enthusiasm needed to carry them to political implementation. Consider, for example, the main recommendations of the National Commission on Energy Policy. They include expanding world oil production, as well as eliminating or reducing sanctions on human rights violators with supplies of oil; building an Alaskan gas pipeline; and convincing Americans that liquefied natural gas terminals can be made safe, as can nuclear power production, radioactive waste disposal, and increasing coal use. The commissioners call for increasing fuel efficiency standards, but only very modestly, while introducing more hybrid vehicles and alternative transportation fuels. Energy efficiency and renewables are encouraged, but with very little change in subsidies, R&D funding, or other real-world supports that would increase them substantially. These proposals also must not harm any business in any way. And, in federal budget jargon, they must be "revenue neutral." They have to pay for themselves through trading carbon credits, whereby reductions in emissions are worth money or credits in a market where they can be purchased by polluters. Such restrictive notions rule out a carbon tax or gasoline taxes, even if Iraq has already busted the nation's bank.

The final result is rather puny. If you add up the costs of all of the recommendations in the National Commission on Energy Policy, they total only $36 billion in spending over the next ten years.[54] This is the amount of profit that the ExxonMobil Corporation made in a single year in 2005 and even more again in 2006 while spending millions to deny global climate change.[55] R&D funds for renewable energy sources recommended by the commission, on the other hand, amount to a mere $360 million per year. The rest of that $36 billion ($3.6 billion a year) is spread out over nuclear power, coal gasification and carbon sequestration, incentives for corporate R&D, and the like.[56] Given the huge amounts that the federal government spends on war and on health care (each over $400 billion annually), and given the risks to the planet, human health, and the nation's security, the National Commission's proposals do not amount to a war on climate change or a clarion call to break the nation's "addiction to oil." Surely, there must be another way.

There is. A number of credible reports challenging the conventional wisdom have now emerged. They indicate that we can get sufficient power for the United States, using renewables and energy efficiencies, to sustain our economy and become energy independent and secure. We can sufficiently reduce carbon emissions so they level off before we reach the dangerous levels of CO_2 in the atmosphere of 450–500 parts per million. But these alternative views are hard to find. There is an inclination in national policy and media circles to take seriously only the reigning consensus. Take for instance, the media coverage, such as it was, of an important energy conference of leading climatologists and energy experts. They met in 2006 and then produced a report called *Tackling Climate Change in the U.S.: Potential Carbon Emissions Reductions from Energy Efficiency and Renewable Energy by 2030.*[57] The *New York Times* environment reporter, Andrew Revkin, for example, used his story on the Annual Solar Conference in Denver to downplay the possibilities and highlight the difficulties. The American Solar Energy Society (ASES) had purposely invited respected, independent experts to address climate solutions. But when the conference chair, Charles Kutscher, announced "Houston, we have a solution!" Revkin scoffed, "Hold the applause."[58] He also stressed, without explanation, that government and business investments in renewables have actually fallen over the past twenty years. Perhaps he was hoping to stimulate them. But his dismissal of renewables as a realistic solution without examining them thoroughly is fairly typical of conventional reportage.

When the report that ensues from a conference such as this starts off with NASA's James Hansen describing how dire things are and then goes on to say: "Energy efficiency and renewable energy technologies have the potential to provide most, if not all, of the U.S. carbon emissions reductions that will be needed to help limit the atmospheric concentration of carbon dioxide to 450 to 500 ppm," we should pay far more attention.[59] The presenters and authors of the report are all volunteers. Most come from staid government agencies like the National Renewables Laboratory at Los Alamos. None can be described as wild-eyed radicals or tree huggers. The nine contributors were asked to calculate how much energy could be gotten and carbon reductions achieved by 2030 from the energy sector, such as wind or solar, in which they were expert. They were told to use reasonable, not utopian, assumptions about technology. When totaled up, the energy produced met U.S. needs while reducing emissions by between 1,000 and 1,400 megatons of carbon per year. These reductions, achievable with an aggressive push for renewables, are then on track to achieve 60–80 percent reductions in carbon by 2050. That is the amount widely recognized as needed to stave off dangerous climate change.[60]

When the Congress changed hands in 2007, the report from the American Solar Energy Society was released to Congress and the public by the Sierra Club, ASES, James Hansen, Representatives Henry Waxman (D-CA) and Chris Shays (R-CT), and Senator Jeff Bingaman (D-NM). It is now the main climate strategy of the Sierra Club, according to its executive director, Carl Pope.[61] But it is still far from the conventional wisdom.

When Chris Flavin, president of the Worldwatch Institute returned to my class in 2007 and 2008, his information was even more compelling. The trends in renewables he had sketched so clearly in his previous visit were continuing to accelerate. Chris looked less the prophet and more someone with a prescient prognosis. The Worldwatch Institute had teamed up with the Center for American Progress and issued a report on the future of energy and renewables called *American Energy: The Renewable Path to Energy Security.* It, too, analyzes the capacity and trends of various renewables; its conclusions are similar to those embraced by ASES and the Sierra Club. A new, secure, and prosperous energy future that also reduces CO_2 emissions can be built on renewables. California is already getting 31 percent of its energy from renewable sources. A similar burst of change is possible at the federal level if citizens, organizations, and businesses band together to push for it. American Energy also contains a link for citizens to get involved and sign a petition. Its policy recommendations are supported by a network of more than two hundred organizations from environmental NGOs to economic powerhouses.[62]

Other analyses of this sort deserve more consideration, too. Many commentators have pointed out that the United States cannot solve the climate change problem alone. Emissions are rising in the developing world. As you read this, China will have passed the United States on an annual basis as the world leader in carbon emissions (though clearly not on a per capita basis, or as a cumulative contributor to overall CO_2 concentrations). Although my focus is on the United States, international studies about renewable energy that include U.S. data and scenarios draw similar conclusions. Greenpeace International and the European Renewable Energy Commission (EREC), for example, teamed up on an expert review and report on energy and renewables called *The Energy (R) Evolution: A Blueprint for Solving Global Warming.* Based on meticulous research, it uses independent experts and analysis by the German Aerospace Centre. Its findings indicate that in the United States and worldwide—given political will, leadership, and investment—we can move to a safe, renewable energy future. Their startling, overall conclusion? "The reserves of renewable energy that are technically accessible globally are large enough to provide about six times more power than the world currently consumes—forever."[63]

With the election of 2008 and beyond, no one can say that we do not have options, working models, and plans that could, given sufficient political impetus and leadership, quickly turn around the U.S. economy and its carbon emissions.[64] This change, like that from the horse-drawn carriage to the Corvette, can happen far faster than most people or politicians have realized during the dark, do-nothing days that, like smog, settled over the United States at the dawn of the twenty-first century.

9

Creating Hope: What You Can Do

I want to end as I began. This book is an act of faith. At the outset, I promised you hope in a few hundred pages. Whether you have stuck with me thus far, or have skipped to the back of the book as I often do, or are idly flipping through the pages in your local bookstore, it's time to close the deal in this chapter and the next. The faith part rests on you and me. It is remarkably refreshing and empowering to think that what you or I actually do or do not do makes a difference in the world. And it is scary. It means you and I are *responsible* and we ought to do *something*. Noah and his wife may have gotten some sort of signal from God, but let us not forget. *They* built the ark and herded aboard that menagerie. *They* took action. If you remember other major motifs in biblical lore, many of the heroes were either quite reluctant or living the good life when the lightning of change struck. A personal favorite is Moses. He wades through the receding tides of the Red Sea, outruns the chariots, and delivers his people to freedom. But when he is first called by Yahweh, his response is pretty much like yours and mine: "Why me, O Lord?" Moses has been leading the good life in the palace amidst opulence and slaves. And now he is going to have to change. That is the scary part. And it is the hopeful part. History and the headlines are full of amazing changes, of people who decided one day that they would do things differently, that they would speak out, or refuse to move to the back of the bus.

Most of us don't like change much. So I want to offer two things before I close. I hope to convince you that even the smallest and simplest of changes, of actions, can make a difference. And I hope to convince you that as you take small steps, they will be most effective, most helpful if you join in with others, in a community, an organization, an institution that shares your goals. Moses did not say: "Let *me* go!" He said "Let my *people* go!" The same story is repeated in many forms throughout history from Mt. Sinai to Selma. Today, we are faced with a dismal future for all of God's children unless, together, we start heading for the Promised Land. Global warming will require new ways of doing things from each of us. They can be small or, in some cases, they will be wonderfully, extravagantly, large. Where to start? Can it possibly make any difference? Good questions. But I am fairly sure that if Noah and his wife just worried whether or not the ark would float, or which direction to head, or how many animals it could hold, or whether they would survive at all, we would not be here now.

I prefer my own routines as much as the next guy. Get new light bulbs? A different car? Change political parties? Go to a rally? Talk to neighbors about . . . climate change? Hey, I'm a guy who has been eating Cheerios since they sponsored the Lone Ranger in the 1950s! I vote. I pay taxes. Let Al Gore do it! All this could lead to arguments in the family, at work, and even harder stuff. A different house, job, career? Look. I'd like to help, but I have stuff to do, okay? You should have sensed by now that I am a reluctant activist. I like people. I don't really like conflict or arguing. My sense of outrage is sometimes just too tepid. Along the way, I have had plenty of doubts that any of it made any difference at all.

And yet, our current story, our own history, is already filled with examples of change and influence, large and small. Take Eban Goodstein, the founder of Focus the Nation. Imagine. An economics professor has taken off from a comfy college job for more than two years to crisscross the country training anyone who will listen how to talk about climate change. And then he simply decides that there *will* be one thousand college teach-ins! Lois Gibbs refuses to take no for an answer when told that chemicals are not the problem in Love Canal. Sandra Steingraber gets cancer in her youth and turns it into a lifetime of writing and speaking on the environmental causes of cancer.

Or take Laurie David. Talk about the good life. When she was married to comedian and producer Larry David, she surely could have continued with the satisfied set. David had worked her way up from copywriter for a car dealership in Cincinnati, "writing snappy lines for the Dodge Dart!"[1] Finally she got a job as a researcher for David Letterman. Then she represented comedians of all sorts. After marrying Larry David, of *Seinfeld* fame, she describes her light-bulb moment. Home with their first child, Laurie says she was frustrated with motherhood. "The baby had colic. Larry was on the soundstage seven days a week, my career was on hold—all my friends worked. I was isolated and scared."[2] As she spent her time pushing a stroller around the neighborhood and feeling protective of her baby, David kept noticing all the huge SUVs roaring by. Then, miraculously enough, she walked into a bookstore and happened upon *New York Times* reporter Keith Bradsher's book *High and Mighty.*[3] In a parting of the seas, she discovers it's about SUVs and how their low-mileage ratings and high carbon emissions are harming the nation and the planet! For Laurie David, the lightning bolts of old, the burning bushes, the thunderous mountaintop voices were transformed into a book. She was on the road to Dalton's bookstore, not Damascus. Yet, the revelation is just as clear. A distant voice has told Laurie David that her child and everyone else's *is* in danger from SUVs and from far, far more. This mother, like so many before her, takes action. Keeping the stroller on the sidewalk, safe from SUVs, reading a book about pollution and climate, is *not* enough. David continued to read more and more books, learned more about the environment, and eventually joined the board of the Natural Resources Defense Council (NRDC) headquartered in New York City. There, she turned to John Adams, the president of NRDC, for advice. Adams, who for years chaired the Green Group, is one of the most respected leaders of the environmental movement. His advice was simple and straightforward.

Focus on one thing. Focus on climate change. It is the most important, the biggest issue for us all. And so David created the Stop Global Warming Virtual March. After she explained to fellow NRDC board member Robert Kennedy Jr., a fiery speaker, organizer, and radio host, what a virtual march means (people sign up, take action, and do things via the Web—all across the nation) change was born. What the virtual march does, and continues to do, is allow people to take a simple first step, to get engaged, to do something. David has gone on to learn how to handle policy, speeches, TV shows, and to write her own first book, *The Solution Is You! An Activist's Guide.* It tells her story. It also has a hilarious introduction by Larry David that is worth the price of admission. But best of all, I think, is Laurie's sense that small things *do* matter. "Start small and grow. Start in your house, then, move to your school, your book club, your gym, your church, your temple, your city. Start with your friends and family. Reach out to people who don't agree with you."[4] I agree. As you know, for some time I thought, or perhaps feared, that if you only changed the light bulbs in your house, you might think that is enough. It has taken me until recently to change my own light bulbs, to recognize that one small step leads to another, that not everyone is drawn to high policy or international treaties, or has the opportunity to influence them. But as you buy your first compact fluorescent lights, I hope you will begin to connect that step to the 500 pounds of coal that will be saved, then on to the miners and scarred landscape, and finally to broader, bolder action that will lead to an end to coal and oil and gas.

Home and Hearth

Let's look at what you can do at home. A light bulb burns out. Oddly enough, in a country that has gone to the moon, it is exactly the same invention given to us by Thomas Alva Edison well over a hundred years ago. I have an antique one in my collection from my father-in-law's hundred-year-old detached garage. Guess what? Slightly different shape. Same incandescent light bulb that screws in. Same filament. Same everything! A great invention in its day, the regular incandescent bulb gives off 90 percent heat and only 10 percent light from the electrical energy it uses. The power needed, of course, is measured in watts—40, 60, 100, and so on.

When you buy your first compact fluorescent light bulb with the soon-to-be-familiar coil shape, it will give off the same amount of light as your old one, while using only a quarter the power. That means that a 25-watt compact fluorescent bulb (CFL) gives off the same light as your old, standard 100-watt Thomas Edison model. But it will account for only one-fourth of the CO_2 emitted from the coal-fired utility that likely provides your electricity. I have mentioned that I was impressed by the 500 pounds of coal that environmental activists showed me would be saved over the lifetime of just one changed bulb. That small act matters. If multiplied across the nation, it would make a huge difference. According to Jeffrey Langholz and Kelly Turner in their excellent collection of practical tips in *You Can Prevent Global Warming,* if every household in the United States replaced its next burnt-out light bulb with a CFL, more than 13 billion pounds of CO_2 emissions would be prevented. That's the equivalent of taking 1.2 million cars off the road for an entire year.[5]

But the big electrical usage and, hence, CO_2 and pollution production, is from major appliances—your refrigerator, washer and dryer, water heater, and such. The refrigerator uses about 13 percent of the energy in most homes and costs about $120 a year in electricity. That's why in the 1990s, my wife and I bought the most efficient model we could find. But as it turns out, while writing this book, I checked and realized that our model is more than ten years old; brand new ones use only half as much energy as my "efficient" model. Ordinarily, I would think of a new one as a major extravagance. But not when I realize that I will save about $60 a year for the next dozen or so years—while cutting back on emissions. In the meantime, I have realized that my penchant for keeping things cold should probably be turned down a notch or two. When I read that if every household in America turned up the temperature in its refrigerator by just 1°F, we "would prevent almost three million tons of carbon dioxide from entering the atmosphere—every year," I ran to the fridge and twirled that dial.[6]

But we could have even better appliances if politics did not prevent it. I have talked about the good work of many major environmental groups. Greenpeace has been known mostly for concern about whales and seals. But it has been campaigning for eons against chlorine products, toxic chemicals, and climate change. When in 1995 the government moved to get rid of CFCs (chlorofluorocarbons) in refrigerators because they eventually seep out and destroy the ozone layer, the replacement choice was hydrochlorofluorocarbons (HCFCs). They are safer for the ozone layer. But they are potent greenhouse gases. Greenpeace worked for an environmentally safe refrigerator and came up with the Greenfreeze, a model that is cheaper to manufacture and is 38 percent more energy efficient than a comparable one using HCFCs.[7] The only problem is, as of 2008, you cannot buy one in the United States. The Greenfreeze has become the most popular refrigerator in Europe. But our manufacturers simply will not make them. So when you upgrade, or until you do, a first small step might also be to write your congressional representative and ask that the government mandate an immediate phase-in of Greenfreeze refrigerator technology.

The same pattern holds for most appliances in your house. Invest in more efficient, greener ones. I just got a much smaller, more efficient gas furnace after living for years with a huge, old, rumbling, inefficient one that came with my 1941 suburban brick home. It was a pretty big investment, but I wish I had switched long ago. Now the bills are lower, the house toastier on lower settings, and the CO_2 emissions are way down. Then comes the water heater. We got a more efficient model. But we still have not wrapped the darn thing in insulation. That alone would save another 10 percent of energy, costs, and CO_2. But short of that, I only recently realized that simply turning down the water temperature to 120°F instead of 140°F (or to the energy-saving setting if there is one), can save another 10 percent in energy use. If only half the households in the United States turned down their water heaters by just 10°F, the atmosphere would be spared 239 million tons of CO_2 emissions per year.[8]

Doing such things and urging friends and neighbors to do the same adds up pretty quickly. David Steinman, best known for promoting healthy, organic foods, reports in his latest book, *Safe Trip to Eden: Ten Steps to Save Planet Earth from the Global Warming Meltdown,* that even toilet paper matters. If each household in the United States changed from regular toilet paper to a 100 percent recycled, non-chlorine-bleached brand like Seventh Generation, that would save 170,317 tons of greenhouse gas emissions for *each* twelve-pack used. Same for paper towels. One three-pack of towels per home would save 63,000 tons. The switch of just *one* 500-pack of paper napkins would stop the release of 120,930 tons.[9]

Remember, too, that recycling everything you use not only cuts out waste to the dump, it saves energy and CO_2 emissions. When you are asked about paper or plastic at the supermarket, the real answer is reusable shopping bags. Most plastic or paper bags do not get recycled. Even when they do, they eventually get discarded somewhere. And then a new one has to be manufactured, shipped, and stored. It all takes energy. It all creates CO_2 emissions. The same is true for every product in your house. Recycle, reuse, rent, and return.

Okay. For now, we've got the inside of the house under control. Outside is a little simpler. A yard and a car or two or three, right? (Big-city dwellers, move directly to the final chapter on political action. You're already skipping the cars, using the subway, walking to work, right? I will get to your apartment or condo in a minute.) One of the easiest and more important things anybody with a yard can do is plant trees. I know that Ronald Reagan told us that they are killers that give off dioxin. But trees really are good for us. When I moved into my Bethesda home in July 1987, it was somewhere near 100°F and humid. There was no real shade. It was pretty much all lawn, except for azaleas along the foundation and the back fence; a gorgeous old crepe myrtle; a few rhododendrons, hollies, and lilacs near the edges; and one grand old southern magnolia that made me think of Trent Lott. Twenty years later, I have two towering tulip poplars, two midsize oaks just beginning to rain down acorns, several good-size dogwoods, a large Kwanzan cherry, a weeping cherry, and assorted shrubs that the birds and butterflies seem to like. In one spot, it is so shady that I have a woodland wildflower garden with rocks, moss, trillium, jack-in-the-pulpit, false Solomon's seal, the works. It's where the hammock is. And that's not counting the sweet gum tree cut down by a neighbor or the birch and mountain ash that before their final demise provided woodpecker holes for wren and chickadee nests.

As you know from chapter 2, the days have only gotten hotter in Maryland since we moved in twenty years ago. But I have been a much cooler, happier camper thanks to my trees. My air-conditioning season is far shorter, bills lower. I can set the temperature higher, pay less, and emit far less CO_2. This is from a guy who sympathized with Richard Nixon while he had the White House fireplace and air-conditioning going simultaneously. But if you multiply my trees by 100 million households, the United States would have fairly massive reforestation. That's what ecologists call a "sink" for soaking up CO_2. Ultimately, trees, which sequester carbon, die off and release their stored CO_2. They are not a permanent solution to global warming. But reforestation offers carbon sinks for a very long time. Cutting down trees, whether in your neighborhood, the Pacific Northwest, or the Brazilian rain forest will only make climate change worse. Close to home, the big benefit from trees is cooling. Trees not only absorb CO_2 through photosynthesis, they also give off water through their leaves from transpiration. Transpiration cools the air around them as the water is evaporated. Near the average tree it is, in fact, 9°F cooler than the surrounding air. And Department of Energy studies have shown that tree-shaded neighborhoods are, on average, 6°F cooler than treeless ones. Strategically planting just three trees around your house, according to the Department of Energy, can reduce your air-conditioning bills by as much as 40 percent.[10]

Planting trees is easy. In my day, I have let the squirrels do it; I have started with seedlings and saplings; and I have had larger trees "installed" by pros and by city workers. As we saw in chapter 6, a number of cities are now creating municipal goals for tree planting to help combat global climate change. Many organizations are helping too. Tree-planting programs and advice of all kinds come from the National Arbor Day Foundation and American Forests, two nonprofit groups that promote tree planting and reforestation. Their partners include corporations like Coca Cola and Anheuser-Busch.[11] They pay for planting trees to offset the CO_2 produced in their operations. You can get help with tree planting or make gifts or get involved with either the National Arbor Day Foundation or American Forests. It's one of those small steps that also will help promote wider shifts in the American landscape, urban planning, and carbon emissions. In the face of global warming, it makes no sense, in my view, to scoff at merely planting a tree. I promise, it will not *prevent* even stronger measures. It will, in fact, lead to them.

On to our cars. Mine is parked outside beneath my front oak and tulip poplar, and one neighbor's giant, shading beech tree. In autumn, it gets bombarded with acorns, the driveway is littered with beech nut pods, and huge tulip-shaped poplar leaves cling to the windshield and catch in my wipers. But it could be worse. I used to have three cars. They could be battered by the squirrels and coated with leaves, too. My reduced car ownership is not really a virtue, however. My daughters have grown up and moved out. As young professionals, they live in cities and exist mainly through walking, a few cabs, and public transportation. My 2000 Honda gets relatively good mileage. But again, what my wife and I call "the new car," is eight years old. I should be ready for a hybrid, right? I am. So much has been said and written about cars and trucks and SUVs, that, suffice to say, whatever you are driving now, your next car or van or SUV should get at least double the mileage. In the meantime, you can drive what you have got much more sensibly. I sit on an environmental group's board with a former EPA and State Department official. He's pretty old school and has not traded in his large sedan. But he has let us know, in charmingly genteel tones, that he keeps his tires inflated, drives at modest speeds, and keeps his vehicle tuned. That's how he manages to match the mileage of a conventional Honda or Toyota. I wrote off such keep-your-tires-inflated stuff when I heard it as a platitude from President Bush. But now I am taking it seriously.

Better still, when possible, is not driving at all. My wife and I can get to work or to National Airport by commuter bus and subway. Washington is a great walking city. But we're waiting for the Metro to get to Dulles Airport (you can only drive) and there are just too many places in the burbs that you cannot reach without a car. So we use one, call cabs, and rent a car when all else fails. I keep asking for hybrids (rarely available, but an option in a few places); refusing upgrades to larger, low-mileage cars; and generally trying to let car rental companies know that they should indeed try harder. The first signs of rent-a-car greening are already appearing, at least in terms of carbon offsets and corporate giving. Enterprise Rent-a-Car, the nation's largest, has announced a partnership with American Forests and the U.S.

Forest Service of 50 million trees to be planted over the next fifty years. That's a million trees a year, or the equivalent of a new Central Park in New York City, with its twenty-five thousand trees, every ten days for the next half century. Tree-planting programs will be selected each year by American Forests. Each million trees, when mature, will absorb about 15,000 tons of CO_2 or three-quarters of a million tons of CO_2 overall. The program celebrates Enterprise's fiftieth anniversary in 2007 and is the equivalent of a $50 million-grant to the nonprofit American Forests.[12] There are also options short of renting a car for the whole day. As I write, I am looking at a brochure from Flexcar, a simple idea pioneered in Europe (I hate it when Europeans get ahead and get credit!), where you can rent a high-mileage car for just a few hours. Then you drop it off at a number of convenient locations, usually near a subway. So far, Flexcar is only available in Washington, D.C.; Los Angeles; Portland, Oregon; San Diego; and Seattle. Zipcar is a similar alternative, and many drivers join both to increase the locations they can use. But it is a simple idea whose time has come.[13]

Now that we've got your house, yard, and car travel greened up a bit (and sent a few e-mails, joined a couple of groups, and made some contributions along the way), let's think about further steps. I am afraid to admit this, but I know people who live in houses that actually *create* energy. They sell it back to power companies. And I know people who only buy their electricity from green power sources, that is, electricity made from renewables like wind, solar, and hydropower. But I have yet to do it. Same excuses as before. So let's do this together. Many utilities in the United States now offer some form of green power.

Things certainly do get harder, I know, if you want to really get down to zero emissions or even produce electric power from your home. What I have just described as some basic steps in a standard house will make a real difference. They will reduce greenhouse gases by anywhere from a quarter to a half. But now let's suppose, that like many Americans, you are going to move or really remodel, or even build a new house. Here is where, if you are very handy, have sufficient funds, or just make energy independence a very high priority, you can make major inroads. First, I would recommend picking or building a home as close as possible to urban centers, mass transportation, and, if you work outside your home, to your workplace or office. That way, you can take commuter train, subway, light rail (remember those clanging trolleys?), or even bike, walk, or jog. My dad, a Wall Street man and no eco-freak, rode the Long Island Railroad every day to the Brooklyn subway and on to the financial district where he walked the rest of the way to his office. Carbon emissions? Ecological footprint from this good-natured conservative? Near zero. If you are moving, try sticking closer to the city, a smaller house, smaller and fewer cars, less commuting by car, and an occasional bike ride. You will be happier and healthier for it, and the sprawl that we looked at in the last chapter might just slacken a bit.

Now for your new house. A green architect can help design a place that uses the land so that you are sheltered from prevailing winds, employs the earth to cool and avoid the sun, draws on geothermal heat to warm and cool the house, and has solar panels to provide electricity (supplemented by your own windmill—now being manufactured in household sizes, as well as the larger turbines like the one at Carleton College that graces the dust jacket of this book).

As with most things, there are good, handy guides for all this. The ones I like do not provoke guilt nor do they beat you over the head. Dan Chiras, for example, the author of *Solar Homes,* now out of print, has a newer offering called *The Homeowner's Guide to Renewable Energy.*[14] In it, he tells how he fell in love with renewables in a most unlikely spot—a parking lot. He was visiting Arches National Park in the summer of 1977 on a broiling hot day when he spotted a single, solar electric module. "This amazing little device cranked out electricity to power a small fan. Park officials had attached small streamers to the fan to dramatize the effect."[15] Chiras's interest and "affection" for this little installation (and there is a picture of it) was stirred because he had been studying the nasty effects of generating electricity from coal. But it was, I think, his imagining a brighter, better future that moved him. Chiras remembers marveling at the fact that "there were no toxic emissions, no mines, no heaps of slag and ash to get rid of. . . . This simple, reliable little device with no moving parts was cranking out electrical energy and sending those streamers and my heart into paroxysms of delight."[16]

But the most interesting aspect of the story is to follow Chiras's progression from simple to stately solar living. He may have started long before you and me, but he started small. He merely retrofitted his first home with solar panels for hot water, insulated the attic, and added a greenhouse on the south side. Next, when he moved, he bought a "passive" solar house. "Passive" is a bit of jargon that simply means using natural sunlight to warm a dwelling by designing and locating it to achieve that goal to maximum effect. Chiras also retrofitted that house. Only in 1995 did he make the big leap, building a super-efficient solar home from scratch. He now has *no* monthly electric bills, as all his energy comes from photovoltaic panels and a small wind generator. And, of course, the house was designed to passively make best use of the sun with south-facing windows, and it is cooled naturally, using earth sheltering, a great deal of insulation, and energy efficient windows.[17]

Even advanced solar homes are changing and improving rapidly. Architects are already designing and offering totally self-sufficient homes that, in effect, provide their own climate. This is especially important in areas that are subject to weather extremes, as in North Carolina, where hurricanes, tornadoes, and ice storms can knock out power from utilities for weeks. Builder and architect Michael Sykes of Wake Forest has been building and improving homes that heat and cool themselves since the 1980s. Business is likely to keep improving at his Enertia Building Systems, which offers thirty predesigned home models in a catalog, as well as custom designs. Each features rooftop solar power, and none, given their design, has a furnace or heat pumps. The building material is composed of strong, engineered, glued layers of 100 percent renewable wood. It is five times stronger than conventional two-by-four wood construction and designed to resist flying apart or loosening during a strong storm. Enertia houses can also withstand complete submersion in floods for several days, unlike frame houses. And because there is very high carbon content in the walls, the house actually sequesters tons of CO_2—making it more than carbon neutral. According to Sykes, "building and living in an Enertia house is like taking 50 cars off the road." The houses also feature greenhouse atriums that can produce enough food, year round, for a family of four.[18]

I've spared you the methane converters, dry urinals, and other interesting stuff that can and will come with the energy-free homes of the future. But I am aware that much of what every resource on energy efficient housing talks about is for middle-class homeowners. They do make up about 60 percent of American families—depending on your definition. But I recall one trip I took to New Mexico where two wonderful, middle-class organizers from PSR had brought together a very diverse crowd. They excitedly introduced a designer of solar homes. At least two people at the table where I sat said spontaneously: "In my community, people don't even have houses."

That's why, sooner or later, we will all need to take the simple step of urging local, state, and national officeholders to build more low-cost, subsidized housing and to make it green. Or as Laurie David has done, you can get involved with or simply financially support a group like NRDC that is working to create environmentally friendly or green housing projects nationwide.

Around Town

Once you are housed, shaded, and ready to hop onto mass transit or walk to the nearest store, it's time to shop, get some food, and head out into the community. Because I like to eat and am not quite ready to grow all my own food, let's go to the supermarket (or the local co-op if you have one). Along the way, we may also stop at one of the local farmers' markets or even a roadside stand where you can still get locally grown produce. Which is the best choice? I have not been able to get everything I need at a single source, no matter how hard I try. But the principles are basically clear. Perhaps the most important is to buy locally grown and organic food. Authors Barbara Kingsolver and Michael Pollan have each described powerfully the crucial differences between our current factory-style food production and delivery systems versus gentler growing and gardening methods at or close to home. At present—after energy-intensive pesticide drenching, production, processing and packaging—before reaching your table, the average American meal travels 1,200 miles by truck, ship, or plane. If just once a month, you and I and only one hundred thousand other American consumers simply switched to buying locally grown food (or grew some of our own) instead of getting the flown-in stuff, we would collectively prevent 3,000 tons of CO_2 from being pumped into the atmosphere.[19]

In addition to asking for locally grown food (or just looking for the sign, given that more and more stores offer it), you will want to go organic as well. Like a lot of Americans, I avoided the stuff at first: shriveled apples, browning bananas, tiny distorted carrots—you know, the early 1970s servings I was forced to eat along with homemade bread that, when hefted, felt heavy enough to have been grown on Jupiter. All that has changed. I have moved steadily toward organic produce, eggs, dairy, even chickens! In this later phase, I associated organic foods with reductions in dangerous herbicides and pesticides that especially harm farmers and farm workers, the residents in rural areas whose water gets contaminated, and us city dwellers who, over time, get too many chemicals in trace form. Results? Cancer, endocrine disruption, and a host of health problems that follow. At present, according to the EPA, two-thirds of American farmers spray one billion pounds of pesticides per year onto the fields where our food is grown. Improving health here in the United States is reason enough to buy organic.[20]

But only fairly recently have I associated organic farming and gardening with energy savings and global climate change. I had been opposed to chemical fertilizers for health reasons. But I needed to rethink about them as *petroleum* fertilizers and remember that those agricultural-looking fertilizer bags come from *factories*. By the time natural gas, for example, is drilled, delivered, and processed into fertilizer at a huge plant, using vast amounts of electricity from a coal-fired utility, the CO_2 emissions are astounding. Then it must be delivered to the farm and spread or sprayed onto the fields. None of this occurs with organic farming.

Not only are the farmers and workers now healthier, we are too. Organic vegetables, for example, are not only free of pesticides, they are actually more nutritious. They contain, on average, 30 percent more magnesium, 27 percent more vitamin C, and 21 percent more iron than those grown conventionally.[21] But the health of the planet and the climate is improved, too, by good old carbon sequestration. The earth's soil already holds twice as much CO_2 as the atmosphere. It is stored by microorganisms that live there. When pesticides are used, beneficial organisms are also killed off, releasing their CO_2 into the air. As the soil gets less fertile, more and more petroleum fertilizer is required. When the bodies of these microorganisms rot, they give off methane, a potent greenhouse gas. And speaking of methane, before we leave food, a word about meat. I have not managed a totally vegetarian diet. But like you, I have cut way back on meat, especially beef. If you and I eat less meat, and only organic meat when we do, we will also cut down on the production of methane from herds and herds of cattle. You knew this was coming, but cows suffer from flatulence and burping and they poop a lot. All that gives off methane. At this writing, the world's 1.3 billion cows give off 75 million tons of methane—and methane is five times more powerful as a greenhouse gas than CO_2. Just in the United States, livestock of various kinds eat 70 percent of all the corn, wheat, and grain we grow and cover 35 percent of our land area.[22]

Then there is the electricity needed to make, say, hamburger. The feeding, slaughtering, packaging, and transport of the stuff contained in just four Quarter-Pounders, or the carefully wrapped pound of lean ground sirloin you bring home, takes the equivalent of a gallon of gas. Each year, the 260 pounds of hamburger and steak eaten by just one average American gives off about 1.5 tons of CO_2. Again, if you and I are joined by only one hundred thousand other Americans in reducing the beef we eat each year by just 25 percent, it will prevent 1,800 tons of carbon from being released.[23]

Now that we're out in the community, one of the simplest things you can do, now that you have already changed the lights and appliances at home, is to change them at your church, synagogue, or temple. (For secular readers, you may substitute local library or school.) Although, like other nonprofits, many churches struggle to meet budgets, the same principles of long-range investment and savings, as well as stewardship, apply. The churches and synagogues in my area are ablaze with banners to "Save Darfur!" They support soup kitchens and housing for the homeless, literacy programs, and more. And, after capital campaigns, many have remodeled so that there is increased handicapped access outside and elevators within. But the move toward saving energy and the planet is really just beginning. You can make a big difference here pretty easily. At a minimum, you can pester the pastor and the trustees or elders to take the exact same steps you did at home. Lights, appliances, more efficient cooling and heating, lowering the thermostat, insulation, the works. Or, if you really are convinced, you can volunteer to lead the effort yourself or get in touch with Interfaith Power and Light or PennFutures and other groups that can help you green up the House of God. And, given the power of collective buying, they can also get you reduced rates, right down to energy efficient exit lights. Given that most urban churches are large, older, inefficient buildings and most suburban ones were built in the 1950s and in need of serious upgrading, it's a good place to start.

And let's not forget virtue. You don't want to be obnoxious about it, but those I have known who are involved in helping to retrofit or raise funds for green buildings, whether sacred or secular, become passionate proselytizers. One friend, Jon McBride, who is almost retired from the executive search business, beams when you ask him what else he's involved in. The answer? A major green building project at his kids' private school. They have since graduated. But Jon has been consumed, along with the trustees, in learning about architecture, construction, carbon emissions, and more. It may have been just a building committee. But it rightly has made him feel that he is helping not only his school, but the planet as well, while setting a precedent for schools nationwide. Now he is involved in arranging tours and publicity for the place!

The same is true for one of my students at American University, Carl Whitman, one of the best I've had. To be fair, Carl was an administrator at AU (he's now an associate vice president at California State, Stanislaus) in charge of information systems. Silver-haired, a tad reserved, always in a suit, he stood out from my more typical younger students. But when we talked about green architecture, he really lit up. He had been on the committee at American University working on the new green building that will become the home of the School of International Service in 2009, after it is completed by McDonough and Partners.[24] Like Jon McBride, Carl spoke with passion about the building and its features. He printed out drawings and architects' and artists' conceptions for all to see. I have observed the same phenomenon across the country as ordinary people get renewed energy from giving it to others.

One favorite was a talk I heard at Harvard Medical Campus from the associate head of facilities, Peter Stroup. Staff and administrators are too often ignored at universities. And they usually don't get the opportunity to lecture. But I have rarely been as entranced as I was by the passion, clarity, imagination, and innovation coming forth from this man. He talked about how he had used computers to coordinate and time the lights, the heating and cooling, the computers—everything—so it was more energy efficient. It was truly humbling and inspiring. And it was a lesson of the power of an individual. The very first green building at Harvard was the School of Public Health; it was promoted primarily by one forward-thinking professor, John Spengler. Only later did the rest of the facilities follow.[25] The lesson is simple. You could do the same. Local church, school, the college that you or your children attend—all could see rapid change and a different climate starting with the energy of just one person.

Leaving Town

Americans travel a lot, and you may want to see some of the green homes (there are tours) or schools or colleges that I have mentioned. You may even end up, like Laurie David, or others, giving speeches on global climate change and what Americans can do to help prevent it. In any case, you will need a vacation. Obviously, transportation is one of the big carbon emitters in the United States. Globally, air travel makes up 8 percent of global warming emissions. In the United States each year about 600 million passengers fly a total of 70 billion miles for a grand total of 107 million tons of CO_2 emissions. I'm sorry to report it, but if you or I get a new hybrid car, we may save the atmosphere from absorbing about 6,300 pounds of carbon a year. But that amount will be put right back in by just one round-trip flight from New York to San Francisco.[26] As an individual, there are only a few ways to reduce this, because only a handful of more fuel-efficient airplanes are currently being developed.

You can save about half the energy and emissions by going on a train. I have actually done it twice, when my kids were younger. Four airfares were somewhat daunting. Instead of flying, we went by Amtrak to San Francisco and, another time, to Seattle. Each trip was a vacation, not business, so we took the two and a half extra days needed. None of us will forget those journeys and the wonders of seeing the U.S.A. the Amtrak way, rather than the United way, or in a Chevrolet. We met a guy my kids still believe as adults must have been Santa Claus, performed puppet shows, gabbed with strangers who did not live, incredibly enough, in Bethesda, and felt and smelled the glories of the American countryside. We saw hairpin curves and wooded hills in Pennsylvania, hung near open windows at dawn in Iowa, smelling the turned earth and sniffing the hogs. Prairies in Nebraska, wheat fields in Kansas, and, in Wyoming, golden eagles, jack rabbits, cowboys, antelope, and buffalo. (We burst into "Home, Home on the Range!") Then came the Nevada desert, the honky-tonk of Reno, and the forested Sierras as we climbed toward California.

Train travel, especially on shorter hops like Washington to New York is now cheaper, more climate friendly, and more relaxing than the airlines "shuttles," which also require taking a gas-guzzling taxi from LaGuardia into midtown Manhattan. Time difference? With security checks in airports and frequent delays, it may be zero. Or, at worst, you'll spend an extra hour, door-to-door. If one hundred thousand people took the high-speed Acela train to New York from D.C. instead of the shuttle, we would save about 3.5 million pounds of CO_2 per year.[27] Buses work fine, too. They have gotten more comfortable, have videos, and are far, far cheaper. As I write, buses have begun to compete fairly effectively with both trains and planes through special express runs.

But sometimes, you just have to fly. What then? The answer, for now, is carbon offsets or tags that you can buy as an individual. First you go to a Web site such as www. chooseclimate.org or www.carbonfund.org or Al Gore's www.climatecrisis.net/takeaction/carboncalculator. There you can figure out the carbon emissions per passenger of your flight or other travel. (I just did it on chooseclimate.org for a New York to San Francisco flight. I got 323 kilograms per passenger, round-trip. One kilogram is equal to 2.2 pounds, so I'll need 711 pounds of carbon offsets.) You then can buy them from either a nonprofit organization like the CarbonFund, or a company like Climate Partners or TripleE.[28] All will invest your funds in renewable ventures like wind power plants, though the projects vary.

Or you can buy Cool Tags for your travel. These cost just $2 each and are offered by Clif Bar, an organic food company that has been growing rapidly. I first heard about them at a campus climate event from Elysa Hammond, an ecologist who works on strategy and communications for Clif Bar. Hammond is an example of a corporate type who would give any consumer hope. It's not just the organic power bars or even the firm's Cool Tags. As she travels and speaks, she educates and inspires. Armed with a Clif Bar company PowerPoint presentation, she waxes eloquent on carbon emissions, organic farming, and more. Like Chris Flavin talking about the trends in renewable energy, Hammond is quick to point out that organic farming is not just some quaint little hippie idea. It's one of the fastest growing businesses around, with a current growth rate of more than 20 percent per year. Total pounds of organic food sold have gone from 2 million in 2002 to more than 20 million pounds when I heard her, just three-quarters of the way through 2006. If the 160 million acres of corn and soy beans in the United States were farmed organically, Hammond tells her audience to applause, the country would save enough CO_2 emissions to meet 73 percent of its Kyoto carbon reduction target.[29]

Clif Bar also sponsors carbon-neutral concerts and environmental organizations and projects, and it has those Cool Tags. You just click on the logo on its Web site (it says "Start Global Cooling" and shows a windmill and snow-capped mountains—the company likes to accentuate the positive) and you are taken to the Web site of NativeEnergy to get your tag. For each one sold, Clif Bar invests $2 in NativeEnergy's Windbuilders program that helps the Rosebud Sioux Tribe build the St. Francis Wind Farm in South Dakota. Each Cool Tag keeps about 300 pounds of CO_2 out of the atmosphere.

If my calculations are correct, a typical cross-country round trip would take about two CoolTags or a $4 investment.[30] That's less than the house wine or the boxed lunch on board your flight.

You can also now plan your flights and get carbon offsets at the same time thanks to a new partnership between Expedia.com, the online personal travel planner, and TerraPass, a carbon offset group that is an offshoot of the Ford Motor Company. You can buy these in three levels, ranging from $5.99 for 1,000 pounds of CO_2 or 2,200 miles up to $29.99 for 5,000 pounds or the equivalent of 13,000 flying miles. With the larger purchases, you get an Expedia and TerraPass luggage tag that says "Carbon Balanced Flyer."[31]

YOU CAN'T DO IT ALONE

At this point, you have come close to doing what you can as an individual. Let me repeat. If each of us did all of this, it would make a significant difference in reducing U.S. and global emissions of CO_2. Individual actions *do* matter. But they are best combined with local community and organizational efforts. Too often, however, I encounter skepticism about the value of community action from people I meet, even relatively liberal, committed intellectuals; journalists; humanitarians; activists; and religious folk. The individualist tradition in America, the myth of the lone genius, and the growing social isolation and disintegration of groups and neighborhoods described in Robert Putnam's *Bowling Alone* keeps many well-meaning people from joining major organizations or political parties.[32] But whether you hope to influence business, the local municipality, state government, the Congress, or the UN, it's hard to do it alone. If you and I are going to help prevent global climate change, we will have to *join* something and get engaged, at some level, in the political life of the nation.

If anything should be clear from *Hope for a Heated Planet,* it is that the actions of a fairly small number of scientists, environmental groups, and political leaders have brought us to the verge of changes to slow down global warming and build a new, sustainable energy economy. They have been up against an equally small number of powerful corporations and their lobbyists, PR firms, and political hacks—with large amounts of money. Only getting you and me further engaged in collective efforts to change government and corporate policies will finally make the difference. And we will need new laws, lawsuits, and treaties to do that—no matter the exact form they take. Like individual actions, voluntary measures are fine, as far as they go. But they are not enough. In the United States, after some fifteen years of voluntary measures from the Rio treaty of 1992 up until now, our CO_2 emissions have *increased* about 20 percent over the 1990 levels we are supposed to be aiming at.[33] Only a new, expanded climate movement, connected to major environmental groups and other organizations representing millions of us in American life can take the next step. And that movement must be connected—in ways that count—to state, federal, and international policy and politics.

Where could you begin? I would follow John Adams's advice to Laurie David. Focus on climate change. It is the environmental, public health, economic, spiritual, and survival issue that will ultimately dwarf all others. First join just a few of the major organizations that have been working effectively on climate change or clean air. Join especially those that have individual memberships and lobby in Washington and state houses, as well as educate, organize, boycott, or march. The short list ranges from quite activist to nearly establishment groups. Many have local chapters or affiliates with offices and active staff. You can join any of them with the click of a mouse, tapping in your credit card number. In no particular order, these include the Sierra Club and Greenpeace on the activist end, the Union of Concerned Scientists (UCS) and Physicians for Social Responsibility (PSR) for credibility and science, and the Natural Resources Defense Council (NRDC) and Environmental Defense (ED) for more clout and respectability. Then be sure to join the two big environmental political action committees (PACs), the Sierra Club PAC and the League of Conservation Voters (LCV). LCV's political committee has representatives from all the major environmental groups, publishes the major legislative scorecard that Congress and the media pay attention to, and focuses on critical races as epitomized by its now well-known Dirty Dozen campaign. That makes eight clicks and contributions. Be sure to give them your e-mail address and get on their activist alert lists. You will soon be better informed and taking far more effective action than all but a tiny percentage of Americans. Each of these groups cooperates with the other thirty-four or so organizations in the Green Group, and belongs to a number of coalitions, too. They do not, repeat, do *not* duplicate each other's work.

Once you belong, take their suggested action as often as you can. Make it a simple habit, like brushing your teeth. In addition to asking you to send an e-mail or a letter or to vote or to come to a lobby day, they will also help you write letters to the newspaper or to corporations, or they may ask you to become active in a local chapter or club and attend a regional or national meeting. Just do it. And then contribute. Every time you are asked.

This short list should help focus your giving. Aim for the "large donor" levels most groups have. They are called the "Leadership Circle" or the "President's Club" or some such. When you give $1,000 or more (that's 83 bucks a month—dinner for two) you will get far more than the satisfaction of seeing your name listed in the annual report. Such gifts build morale among the staff. They are generally very bright, young, hard-working, unsung heroes of our time. Just twenty such checks are the equivalent of a small foundation grant that may take months and months of work to obtain. Your chosen organization can instantly hire a part-time organizer or buy an ad or turn out thousands of calls to voters during an election. And, you can be sure, you will begin to have a voice in an organization that is actually influencing the public and public policy. Along with key board members and knowledgeable senior staff, like other environmental CEOs, I actually called and visited such donors. And we listen. From contributors I have gotten ideas for development campaigns and state policy initiatives; inside dope on officeholders; free advice on communications, lobbying, and computers; and candid feedback on what my group seemed to be doing well and what not. Although I do not and never did wear Gucci loafers or a Saville Row suit, this is the sort of influence that those folks we love to hate—polluters, plutocrats, influence peddlers—pay *really* large sums to acquire through PR firms and K Street lobbyists.

There are other good organizations out there as well. You should probably join some of them, too. Just don't develop a double Rolodex full. All of the big groups are doing something on climate change, as we saw with the polar bears at Defenders of Wildlife, or the Campus Ecology program at the National Wildlife Federation. NWF does even more than that because, in another example of the influence of a single individual, their CEO, Larry Schweiger, told the NWF board of trustees, when they offered him the job, that he would not take it unless they agreed to do more on climate change. I also like 20/20 Vision, a small, grassroots educational and lobbying group that decided to focus on our nation's energy security. With a campaign called PowerShift, and a remarkably small and efficient staff, it has carried out nationwide events on campuses and in cities, with a simple message:

our dependence on oil not only creates climate change and pollution, it undermines our national security. I cannot name all the other groups out there, but most book buyers like you can afford to give the price of a hardback, $25, to another twenty or so organizations. They will be happy to have you as a paying member and part of their activist network. So tack another $500, or $10 a week, onto your climate mitigation plan.

Becoming More Active and Engaged

If you have gotten this far, you are probably about ready to do even more. Reducing your carbon footprint by lowering energy usage in your home and community is great. So is joining and contributing money, in whatever amount, to environmental and related organizations. But becoming an activist to prevent climate change is not as hard as you think. It can be fun. Your talents and confidence will grow, and you will meet many interesting, even inspiring people. It may not take more time than you or people you know already spend on the gym, the church choir, a bridge club, a book group, or plain old channel surfing.

One of the first, easiest steps—backed up by some solid information and advice from an organization—is communicating with the media. I like to say to those people I meet who ask: "What do you think will happen in the next election?" or "What will President Bush do?" that American political engagement is not a spectator sport. If you just sit back and watch, something you don't like is more likely to happen. Put another way: "If you don't like the headlines, make your own!" A letter to the editor, an op-ed piece, a news item about an event or rally you organized, or coverage of a winning vote on a major bill you lobbied for can get into the papers and onto the news.

The basic principles are pretty simple. First, you've actually got to do it. When my wife and I were in graduate school, the novelist Saul Bellow wrote *Herzog*. The main character penned endless letters to the media that he never sent.[34] I have written and published many a letter and op-ed, but I have a mental stack of thousands of undone, unsent pieces that Herzog would have loved! As with changing your light bulbs, the trick is to do the first one. Then keep at it, and ignore any rejections. To increase your chances, follow the op-ed or letters page, and see what does get in. Adapt as best you can to the style and preferences of your paper. Hook your idea to something that is current and newsworthy or respond to something already in your paper. Keep it short. Avoid rant and invective. And pretend that you are the op-ed or letters page editor. Would you print it? If your media market is small enough or friendly enough, over time, get to know the editors and reporters.

As a member of an environmental organization who becomes active, sooner or later you may be asked to go to an editorial board meeting along with staff experts from your organization. Let the staffer from UCS or wherever cover the hard stuff. You provide a local entrée and common sense thinking about why the paper should write an editorial on climate change or how to reduce the local asthma rate. Or, offer to write an opinion piece yourself. Your temptation will be, of course, to lecture or hector your local editor. Don't. All communications with the media or your representatives are exercises in *persuasion.* Just keep imagining yourself on the receiving end. It will not help if when you leave, the group you have just met with mutters: "Who *were* those asses?"

Some of my favorite PSR activists, busy doctors who cannot spend endless hours in meetings or marching, are well known to their local newspapers. When I visited Portland, Maine; Portland, Oregon; Harrisburg, Pennsylvania; Wichita, Kansas; or Santa Fe, *they* would take me to the editors, the TV producers, the radio talk hosts. In a number of cases, a lone PSR member mostly signed checks and wrote op-eds. Not because they were passive members, but because, like you and me, they lead extremely busy lives. They want to make a difference in the most efficient, effective way possible. So just start in. Start small, probably with a letter to the editor—not to Herzog. And don't expect to be a media star right off. But be warned. You may end up one.

Then, once you're in print, you can move on to lobbying, organizing, and election work. These, in my view, are the most neglected aspects of civic engagement in American society. Most of you reading this book vote fairly regularly, though surveys by the League of Conservation Voters have repeatedly shown that even among members of environmental groups, as many of half may not vote at all. Voting is really the absolute *minimum* that any citizen in a democracy should do. Remember Ben Franklin emerging from the meetings of the Continental Congress? He was approached by a woman outside Independence Hall who asked him excitedly: "Mr. Franklin, what sort of government have you given us?" He replied: "A republic, madam, if you can keep it!" A similar exhortation or warning, from the Bible and quoted by many, is embossed on the Liberty Bell that rallied the Founders and nineteenth-century abolitionists alike. It reads: "Eternal Vigilance is the Price of Liberty."

The First Amendment to the Constitution guaranteed not only freedom of the press and of speech, but the right to "peaceably assemble" and to "petition for a redress of grievances." Assembling and petitioning, in effect, mean organizing, holding rallies, and *lobbying.* The word has come to have some negative connotations, especially in the days of Jack Abramoff. We have seen how powerful the oil and gas lobbies can be. But I have led groups of citizen lobbyists through the corridors of Washington and won votes through what is called public-interest lobbying. It is a fundamental First Amendment right. And it is absolutely necessary if you hope to change any of the laws or policies of this nation.

It, too, can be fun, once you get into it. Sign up for a lobby day or national conference of your favorite environmental or climate change group and come to D.C. You will be amazed—after you get lost a few times and wear out some shoe leather walking the halls of Congress—that you are actually listened to. Typically, you will have been briefed or trained by your group's legislative director, given talking points, background information, a form to write down what you find out, and a map of Capitol Hill and its buildings. Your congressional representative or one of his or her staff will always see you, especially with an advance appointment, if you are from the home district. And I have had many, many members of Congress *beg* me to get people to call, write, or visit. Usually they only hear, noisily, from the opposition or from the well-paid Gucci-wearing lobbyists who actually do roam the hallowed halls.

If all this organizational work, activism, and lobbying fails, what then? The answer is not to give up. It is electoral politics. I am still surprised at the number of people I run into who express disdain for American politics. They will not, or do not know how to, get active in political life and political campaigns. The litany of complaints is familiar and contains elements of truth: there's too much money in politics; corporations really run everything; my vote doesn't matter; the major parties have no fundamental differences; the major parties are too partisan and bitter; the media really tells people what to think. The list is long. But I can assure you that *not* getting engaged in politics will only allow all those forces whom you think run things to *keep* running things. They count on it. It's why most Americans, as measured by polls, can be concerned about global climate change, clean air, and the environment while we have had a Congress that votes regularly to destroy these things.

That has begun to change as of November 2006, when the Democrats broke the conservative Republican stranglehold on power. But it took the engagement and action of many, many people and PACs made up of people like you. They were determined that if they did not like the headlines, they would write their own. Much of the change in Congress in 2006 was fueled by growing opposition to the war in Iraq, but energy independence and climate change, with clean air issues connected to them, were well represented. The LCV and Sierra Club alone spent more than $11 million in 2006 and engaged thousands of volunteers to make calls, knock on front doors, and speak and write about candidates across the nation. Environmental and other nonprofit organizations hold training sessions on how to engage in electoral politics as well as in lobbying. They spend money to make sure that people who care about climate change actually vote, and that candidates who don't, lose.

This is where your activism, involvement, and engagement with serious politics will truly matter. Time is getting very short to deal with climate change. Half-measures will no longer be enough. Our new president will need to lead the world on climate change, to offer vision and hope for a better future. American politics being what it is, you will be offered shades of difference, sound bites, TV ads, and half-baked opinions from pundits and talking heads. You will want to throw up your hands and think about something else. Don't. I mentioned earlier that all your good efforts to save gas and carbon emissions by buying a Prius could be wiped out by a single cross-country plane flight unless you buy carbon credits. Similarly, all your home improvements, local activism, marching, or boycotting could be wiped out if a president refuses to sign new mileage standards for our national auto fleets. Or put positively, a presidential candidate who endorses raising federal fuel economy standards by even 2–3 miles per gallon and who has some sort of record to show that he or she will follow through will save enough oil to replace all that we get from the Middle East.

What this means in practice is that you will want to do some simple but serious research on the 2008 presidential candidates (and Congress, too), as if you were shopping for a new car and looking at *Consumer Reports*.

Check in first with environmental groups. Their only goal is to protect the environment and prevent global climate change. Don't fall for the characterizations you hear and read about a "left-leaning organization" or an environmental "advocacy group." Who do you *really* think cares about and knows more about protecting our climate, ExxonMobil, your evening newscaster, or Environmental Defense? All of the important organizations offer in-depth, well-researched reports, voting records, and up-to-date information on Congress and climate. As I write this book, I do not know who the final candidates will be, so I cannot pick now. But, in chapter 10, I will offer some policy principles and guidance that I hope you will use in 2008, along with information from the environmental groups you have joined and the research you have done. Of course I hope candidates will read *Hope for a Heated Planet* and do exactly as I say. But I am a realist, as well as an idealist. What matters is what *you,* along with your friends and families, your neighbors and networks, and the organizations you join and become active in, actually *do.* That's what determines who will have power and run this great nation of ours. Join a group, vote, lobby, and make sure that you elect officials who share your hopes and dreams. Change a light bulb. Make some headlines.

10

Hope for a Heated Planet: Policy and Politics

My friend Paul Gorman, founder of the National Religious Partnership for the Environment, and I remember quite vividly sitting in the big, high-backed chairs in the White House Cabinet Room, talking about global warming with President Bill Clinton. No group of environmental leaders had ever before met with the president of the United States like this in the inner sanctum of American power. I talked with the president about malaria and the spread of infectious diseases. Others covered a wide range of climate-related policy issues—all of which Clinton seemed to know about. But Gorman had come late. Even though Clinton liked to linger and talk, there was not much time for elaborate speeches. Vice President Gore introduced Gorman and said: "Tell him what you told me before."

Gorman took a breath, dramatically paused, and simply said: "Mr. President, we are messing with God's creation. Global climate change is fundamentally a moral issue, a religious issue. And all the great religions care about all life on Earth. I urge you to keep in mind and remind people of the words of scripture, from Deuteronomy: 'I put before you Life and Death. Therefore, choose life, that you and your children may live.'" Clinton smiled, nodded, and said: "Paul, I'll do that." Then he got a glint in his eye. He looked over at Treasury Secretary Larry Summers, and said: "Larry, how does your shriveled economist's soul like that!" Our group broke up chuckling as Clinton's impatient aides, who had been hanging around the door, finally were able to get him to leave.[1]

I may have been hard on Summers earlier. He could be woefully curt and imperious. Yet he remains an esteemed economist and expert on some things I only dimly understand. It is also clear that underneath an exterior and style that got him into difficulties at Harvard, there is a soul. One of Summers's achievements that should be noted in a long, distinguished "Crimson" history, is his sustainability initiative—Harvard Green Campus.

It now reduces energy usage and carbon emissions throughout the Cambridge campus.[2] This from a man who, by all accounts, raised constant quarrels and quibbles about the short-term costs of saving the planet. I like to think that somehow, beneath the gruff exterior, the policy talk, and the numbers, Larry Summers took action at Harvard because of Deuteronomy, because of the words of Paul Gorman, or maybe those of Bill Clinton, that day in the White House.

It is perhaps a parable for our times. I urge you to understand and learn about global climate change; the destruction of ecosystems and species, the economic impacts, and the immense harm to human life. But foremost, see it as a matter that touches your own faith, your own deepest moral values. Beneath economics and medicine and geophysics and political science, there are values. And choices. Even if they can be muddied by methodology and the paralysis of analysis. Our nation, our families, indeed the world, are now at a crossroads. We must choose. And I believe that to continue to pour millions of tons of CO_2 into the atmosphere, to heat the only planet that we have, to harm our neighbors and those around the world—is just plain wrong.

Now that you are ready, along with me, to make some changes in your own life, and that of your family and community, now that you are becoming more active, it's time to put our renewed energy and commitments to work in the public sphere. We choose candidates to represent us at every level of government. Each matters and has its appropriate sphere of influence—from suburban sprawl, to statewide energy standards, to State Department sustainability negotiations. And we will shortly elect a new Congress and, most important, a new president. You can reach out and communicate with them.

To be effective, you will want to know enough to speak directly about many aspects of global climate change: what it means to your family and our nation; its health, security, and sustainability implications. I hope that this book and many other sources will help you do that. But you do *not* have to be an expert. We are not suffering from a crisis of expertise, or even energy policy. We are suffering from a failure of imagination and of the human heart. Speak to your representatives and candidates directly about their values, their feelings, their faith, *their* choices for the future of our children. Speak, if you will, to their inner Larry Summers, the one that lurks beneath the gruff and grumbling, or grip and grin, exterior.

Ask candidates (or corporate directors or college presidents or county executives or church leaders) whether they think it is right to threaten our children's future by continuing to pump 6 *billion* tons of CO_2 into the atmosphere each year. Ask whether it is right that ExxonMobil, while fighting the facts about climate change, made $36 billion in profits in 2005, $39 billion in 2006, and more than $10 billion, some $1,318 per second, during the third quarter of 2006.[3] Ask, too, if it is right or wrong to fight in wars of choice to protect oil that can and must be replaced. To wax biblical, this process, done often enough—in the right tone and spirit—will soon separate the wheat from the chaff. Paul Gorman's tone is sometimes humble or humorous, but never haughty or full of hokum. Just gently ask: "You are a person of faith and have spoken often this evening about moral issues like abortion and pornography and the plight of the poor. Please tell me what your deepest moral or religious beliefs tell you about whether it is right or wrong to keep up our fossil fuel economy that is causing global climate change. What will *you* do, what steps will you take, to prevent this harm to our children and our future?" There you have it. That is what lurks beneath the numbers and the numbing expertise. The wise citizen, the wise candidate, will address our dilemma directly, without spin, with Deuteronomy in mind.

This approach, of course, risks becoming moralistic or nagging. You may just not be comfortable talking this way in public. I will get to policy and politics in a minute. But as my more prophetic and gloom-and-doom friends remind us, the hour is late—very late. Our confrontation with climate change has been lazy and lax; we are talking about the fate of civilization, of life, not just the price of petroleum. I also ask you to try out different approaches for different occasions, different opportunities. Putting faith first may be best for a personal or private discussion, not a policy debate; best for a living room, not a lecture hall. And it may work best when your congressional representative or candidate is *outside* Washington. The heartfelt, the hortatory, the hellfire approaches all work better as an outside, grassroots, *at home* strategy. As we have seen, inside Washington, you will want to offer some policy ideas, support legislation, push for the very best you can get as a compromise this year. Then come back again and again—and keep at it. The capital is where the laws and policies of our nation are produced. The process *is* long and complicated; at times, contentious and overcautious. But the Founders did not count on perfect people, nor instant results, nor a government where God's will (or our own version of it) would directly and miraculously be chiseled into stone. If we are finally going to get the change we seek, you and I need to translate our beliefs into policy proposals and the language of law. That is the *inside* strategy. With the help of organizations and environmental advocates who speak to policymakers day in and day out, you *can* make a difference.

Gaining Political Clout

I urged you in the previous chapter to get involved with the new climate movement. Personal concern about global warming must become a climate movement, and it must learn to have *political* clout. Public opinion has cared about climate change and the environment for some time. That concern has been growing. You may have already greened your house, bought a hybrid vehicle, gabbed with the neighbors, greeted the Gore movie with glee. Environmental organizations have grown, and new ones have been created. But if there is one failing that many Americans concerned about climate and other causes share, it is that we are not *political* enough. This situation has been changing, too, as in the elections of 2000. Environmentalist and global climate change advocate Al Gore won the popular vote while becoming the first person to do so and not be installed as president since William Tilden beat Rutherford B. Hayes.[4] Environmental organizations and PACs increased their efforts in 2004. The total popular vote for John Kerry, another champion of battling global climate change, increased, as did the number of young Americans who voted.[5] But not by quite enough. The combination of voter challenges and suppression, faulty voting machines, "get out the vote" operations, last-minute campaigning in marginal districts by wartime President George Bush, money spent on attack ads, and all the *mechanics* of politics became, once again, far more important than most Americans realized. Take, for just one instance, the Swift Boat campaign attacks on the medals and heroic actions of John Kerry in Vietnam. I personally asked almost everyone I met if they were aware that John O'Neill, one of the primary proponents of this vicious smear campaign, had been encouraged to make these same allegations, back in Watergate days, by President Richard Nixon and his hatchet man, White House Special Counsel ("I would walk over my grandmother") Charles Colson. According to Nixon, Kerry was too impressive, too articulate, and too credible in his antiwar speeches to be allowed to go on untarred.[6] The lesson here is not that all politics is dirty. It is that involvement in it, vigilance against those who would besmirch open and fair elections, and learning and acting on the *details* of the political process all matter.

There are a couple of simple ways to get involved. First, join the main environmental political action committee (PAC), the League of Conservation Voters (LCV). LCV also maintains a tax-deductible, educational, and charitable arm that, after a section of the Internal Revenue Service code, is called a 501(c)(3). It also has another entity, a 501(c)(4), that can lobby and advocate positions in elections, but cannot raise funds for or work on behalf of political candidates. Join and give to these year round and in nonelection years. From these three LCV organizations, you will be able to get a steady stream of information on issues, legislation, and the voting record and ratings of every member of Congress. You can go to specific training sessions in your area on how to become engaged in election campaigns, lobbying, and more.[7] The same is true of the Sierra Club, the other major environmental and global climate change PAC.[8]

I was shocked, for example, when the *Washington Post* endorsed Maryland Governor Bob Ehrlich for reelection in 2006, saying that he was "good enough" and that "one-party rule" would be bad for the state.[9] Newspaper endorsements do not generally swing elections, but they can set a tone and make a difference in very tight contests. Again, it was only members of the Sierra Club, or similar groups in Maryland, who would know the detailed truth. Although Ehrlich claimed in ads and literature to be for the environment, he had fought tooth and nail against the most important pollution and climate issue in the state. Ehrlich had bitterly opposed efforts by environmentalists and the Democratic legislature in Annapolis to curb power plant emissions. Then, after the bill passed, he turned around, signed it, and took credit during the election season. Fortunately for clean air, people with asthma, and climate change in Maryland, environmental groups campaigned hard (without the concern or coverage of the *Washington Post*), and Ehrlich was beaten in a close race by Baltimore Mayor Martin O'Malley.

The election of 2006 showed what political engagement can do, especially in terms of the war in Iraq, and, to a lesser but significant degree, the environment and public health. Contributions and actions by environmental groups in 2006 made a critical difference in tight races. This was true for the key Senate race of Jon Tester in Montana and the defeat of Rep. Richard Pombo in California, one of the most notorious antienvironmentalists in the 109th Congress. Nationwide, LCV, the Sierra Club, and a few other environmental groups spent more than $3 million (down from the $11 million spent in the presidential year 2004—it's harder to raise money for off-year elections).[10]

Here, then, are some simple suggestions of what you can offer—backed by organizations that work to prevent climate change—to candidates and Congress members who, in an election season, will pay especially close attention.

Develop a Clear Message

Keep in mind the progression of arguments and ideas that we have looked at together. The science is very clear, and it has been so since at least 2001. Human activity, specifically the burning of fossil fuels—oil, gas, coal, and their derivatives—is fouling our air, raising the concentration of CO_2 in the atmosphere to levels unknown in human history, and altering climate worldwide. There are massive ill effects for planetary systems and for people. Using and depending on fossil fuels is undermining our national security. It is skewing our nation's foreign and military policy toward intervention and empire—in the name of oil. Various economists have calculated that we will spend more than $2 *trillion* on the war in Iraq alone. And we continue to allot significant tax dollars—yours and mine—to support our oil habits through subsidies, royalties, incentives, tax breaks, and more.

We must stop, and stop now. Solutions are at hand that can revive American competitiveness, clear the air, and create a sustainable, healthy future. New forms of energy—wind, solar, biomass, and more—are already available. They are growing rapidly. We can now see signs of a shift in climate policy, from initiatives at the state level, to businesses racing to be sustainable, to other nations taking action to cut their carbon emissions. And, most of all, we know there are millions of engaged American citizens and their churches, communities, and organizations working to prevent global warming. Between 2003 and 2006, according to an MIT survey, Americans moved climate change to the top of their environmental concerns. It had been sixth. About three-quarters of our citizens felt the government should do more about global warming and were willing to pay about $21 a month more for electricity in order to help. Americans increasingly see stopping global climate change as an urgent priority.[11] All these are signs of hope. But we need bold leadership and a national commitment to a new economy, a new future. Our elected officials and a government of and for the people can give that leadership. They can stimulate concern and set goals and guidelines. Here, then, are some broad principles and policies for the next Congress and the new president.

PROVIDE A BOLD VISION AND A MAJOR NATIONAL COMMITMENT

On January 20, 2009, the forty-fourth president of the United States will be scrutinized by the entire world. Following years of failed foreign policy and war and a U.S. refusal to lead on global climate change, our new president will have an opportunity unrivalled since John F. Kennedy called on Americans to "pay any price, bear any burden." An inaugural address is the moment to speak as a moral leader, to rally the republic, to provide both vision and values. A clarion call to embrace a new energy economy, to end the era of oil, to combat climate change, and to cherish our children must be the centerpiece.

Shortly afterward, the president will have the opportunity, by tradition and by law, to address in the House of Representatives the entire newly elected 111th Congress. It is a rare chance to quickly and decisively change the climate of indecision and denial about global climate change that has prevailed. The president must then quickly present revisions to the proposed 2009 fiscal year federal budget set forth by the outgoing Bush administration. This offers a telling test of what, with our dollars, will really be done. As with the inaugural and State of the Union addresses, the budget must be bold. Subsidies for the old fossil fuel age must be cut. Business as usual has brought us to the brink of disaster.

With an agenda thus laid out, the president can also take immediate, innovative action through executive orders. These can repeal the worst and install the best of energy and environmental policy throughout the federal government. Carefully chosen and calculated to slash CO_2 emissions, presidential action can ripple throughout the land—much as did President Truman's July 26, 1948, Executive Order 9981 to desegregate our armed forces,[12] or President Kennedy's desegregation directives, backed by sanctions, that included off-base housing near military bases throughout the

Deep South.[13] Then, because all politics is now global as well as local, the president should outline a new international treaty on global climate change that goes far beyond Kyoto and even beyond the initial framework for a new treaty agreed to in Bali at the end of 2007. All this can and should be done within the first one hundred days. It will be a daunting challenge for the next American president and for the American people.

Previous generations of Americans have faced such challenges before—and met them. After Pearl Harbor, the United States mobilized to stop Nazi and Fascist aggression and to fight for freedom. We created an entirely new economy and wartime production that became the basis for unprecedented prosperity in the postwar world. Just look at the speed and size of the shift in military production. Between 1918 and 1933, the United States produced only 35 tanks. After the German invasion of Poland in 1939, the United States produced 309 tanks in 1940. But by war's end in 1945, American factories had turned out 88,430 tanks. The same is true for aircraft. In 1939, the United States produced 5,865 planes. But between 1940 and the Japanese surrender in August 1945, America produced 303,717 aircraft, with dramatically increased power, weight, and speed. Industrial production in the United States doubled between 1939 and 1945, with growth rates of 15 percent per year. Between 1940 and 1942, spending just on factory construction doubled each year from $2 billion in 1940 to about $8.5 billion in 1942![14]

We face an even larger enemy today; it threatens freedom everywhere. To slow and stop the destruction of our planet's climate systems, surely Americans—who pioneered in railroads and mass production, in electronics and the computer age, who have mobilized for war, created the largest university systems in the world, built 42,793 miles of interstate highways, put men and women into space and onto the Moon—can end the age of fossil fuels.

Using current estimates, the United States will spend more than $2 trillion on the Iraq War. We spend only about $20 billion annually for diplomacy and international relations and aid of all kinds. It is a figure that has remained near 1 percent of federal spending since the 1980s. If we are ever to change our foreign and military policies that often focus on fighting for oil and other resources, we must make strong, active diplomacy and development aid hallmarks of America's role in the world.[15] But equally shocking and out of line with serious national priorities is the amount being spent on federal and private research and development for energy technology.

Even as awareness of climate change has been growing, federal spending on energy research has been falling steadily since its peak in 1979, the end of the Carter administration. Then it was $7.7 billion in inflation-adjusted dollars. In President Bush's 2006 energy budget, it was $3 billion. President Bush sought $4.2 billion for 2007, but the amount earmarked for the development of truly clean and healthy renewable energy remained near 1979 adjusted levels, or $1.2 billion. Meanwhile, federally funded medical R&D quadrupled since 1979, and military research increased by more than 260 percent, to stand at $75 billion.[16] Since 2007, budget requests and spending for renewables have remained stagnant, with the president's final FY2009 budget submission still calling for only $1.26 billion for truly clean energy.[17] Most troubling, notes energy expert Daniel Kammen, in *Issues in Science and Technology,* this lack of vision "is damaging the business environment for existing and start-up energy companies. Investments in energy R&D by U.S. companies fell by 50 percent between 1991 and 2003."[18] Kammen, who holds a chair at the University of California, Berkeley, Energy and Resources Group, also testified before the House Government Reform Committee. He said that a number of studies indicate that between five and ten times of the recent amounts of federal spending on energy research will be needed, at a minimum, to stabilize CO_2 concentrations in the atmosphere. Kammen went on to say that the Bush climate change strategic plan (CCSP) to reduce carbon intensity "is seriously flawed." Its goals are far too modest. The Bush plan would actually lead to an *increase* in CO_2 emissions of 12–16 percent. Kammen also indicated that the mandated reductions of the Kyoto Protocol, the 25 percent reductions passed by the California legislature, and the long-term reductions goal sought by Governor Schwarzenegger show a clear path to necessary reductions—the real goal of any serious plan.

Using IPCC data for emissions and savings from energy technologies, Kammen testified that an investment of five to ten times 2006 levels "is not a 'pie in the sky' proposal. It is, in fact, consistent with the scale of previous federal programs." Such an effort "could result in potentially large economic benefits." Even current levels of investment, he noted, have brought down the costs of solar cells by a factor of twenty and wind turbines by a factor of ten. Kammen went on to compare scenarios of increased energy R&D with prior federal R&D efforts such as the Apollo space program, which added $127.4 billion between 1963 and 1972 or Reagan defense R&D, which added $100.3 billion to the federal budget. Even the larger tenfold increase in energy R&D would only add a comparable amount, or $105.4 billion, over ten years.[19]

That the U.S. and other economies can grow—even as CO_2 emissions are reduced and stabilized through massive programs—was underscored at the time of the 2006 U.S. elections. The Stern Commission, an official review by the British government, announced that catastrophic climate change could be avoided and benefits obtained if only about 1 percent of global GNP were invested in CO_2 reduction programs. Media coverage stressed the report's warning of catastrophic harm to the planet and to society if action was not taken. Immense economic losses—up to 25 percent of world GNP in some scenarios—could ensue.[20] Yet the message from Sir Nicholas Stern, head of the Government Economic Service, was actually quite *different.* "The conclusion of the Review is essentially optimistic," he said. "There is still time to avoid the worst impacts of climate change." Stressing economic opportunities and global justice, Stern pointed out that all countries will be affected, "but it is the poor countries that will suffer earliest and most." Spending about 1 percent of global GNP each year, Stern indicated, is quite manageable. People would pay a little more for carbon-intensive goods, but our economies could continue to grow." That's because each ton of CO_2 emissions right now is causing about $85 in damage, costs that are not included for investors in current energy. Emissions trading schemes already in operation since Kyoto "have demonstrated that there are many opportunities to cut emissions for less than $25 a tonne [metric ton]." In short, Stern claimed, "everybody [would be] better off." According to one measure, he said: "the benefits over time of actions to shift the world onto a lower carbon path could be in the order of $2.5 trillion each year. . . ." He went on to remark that "climate change is the greatest market failure the world has seen." To change that, three things, he believes, are essential on a global scale: carbon pricing, through taxation, emissions trading, or regulation; technology policy to drive the development of low-carbon and high-efficiency products; and the removal of barriers to energy efficiency and education of the public.[21]

In a U.S. context, our next president should announce and ask the Congress for a major, long-range commitment to build a new, sustainable economy. The goal should be to cut CO_2 emissions of the United States by at least 20–25 percent in the next decade and by 80 percent before the middle of this century. Those are the reductions the nation's and the world's best scientists tell us we will need to avoid the worst effects of global climate change. Ideally, before this century ends, the United States will need to be producing energy and products with *zero* carbon emissions.

Given the dangers of nuclear proliferation, as witnessed in North Korea and Iran, the president should also recommit the nation to nuclear nonproliferation and nuclear disarmament with a call to eliminate nuclear power. Here, there will be increasing pressure to consider nuclear power as a green fuel, because it does not produce CO_2 at the point of electricity production. But given the continuing vulnerability of such plants and their shipments of nuclear fuel and wastes to terrorist attack or diversion, nuclear power would also negate the strong response needed from the next president to make greater security at home a priority.

Our new president will need popular backing for any massive new program to fight global climate change; only a bold call that makes clear a commitment to the health and well-being of Americans can carry the day. The chief executive should recognize that renewable sources of energy can have tremendous potential appeal—for climate change, for the economy, and as a vision for a truly sustainable future. And there are plenty of solid plans and recipes to draw on. The president needs to rely less on conventional wisdom and more on innovative, yet practical thinking like Chairman Charles F. Kutscher's announcement at Solar 2006, "Houston . . . We Have a Solution!" Listen closely to the experts who agreed that the United States can meet needed CO_2 emissions cuts with renewables and energy efficiency. According to Kutscher, a scientist with the National Renewable Energy Laboratory, 1.3 billion tons of carbon, "greater than what's needed" could be kept out of the atmosphere by 2030. Savings, detailed by experts in each area, can come from solar power, 120 million tons; photovoltaics, 125 million tons; wind power, 300 million tons; biomass, 130 million tons; biofuels, 60 million tons; geothermal methods, 65 million tons; efficient buildings, 200 million tons; and more efficient vehicles, 150 million tons.[22]

Given the memories of difficulties with solar technologies and investments from a generation ago, conventional wisdom still tends to assume that renewables will remain small-time solutions. Only a mix of nuclear plants, "clean" coal, natural gas, and other nonrenewable, or potentially hazardous, technologies will fill the bill. But take, for example, solar power. Greenpeace partnered with the European Photovoltaic Industry Association in a comprehensive report called *Solar Generation.* It argues persuasively that by 2020 solar power, given policy changes and support like that in leading solar nations such as Germany and Japan, could provide electricity to more than 1 billion people and create 2 million jobs, while reducing poverty in the developing world. The solar industry, a tiny fraction of energy production now, is growing by more than 35 percent annually. In the European Union alone, business tops $7.5 billion. Much larger gains in production and carbon reductions are projected by 2040.[23]

If the United States is to free itself from oil dependence, clean its air, improve public health, and cut carbon emissions on the scale necessary, candidates of every sort will need to push for strong renewable portfolio standards and policy and financial changes that move renewables rapidly into the mainstream. This can happen at the local level, too. One source useful for tailoring local calls for renewable energy is from the Apollo Alliance, a national coalition of environmental groups, labor, business, and other constituencies whose president, Jerome Ringo, we met earlier. It has issued *New Energy for Cities: Energy-Saving and Job Creation Policies for Local Governments.* Addressing the time frame of 2025, the Apollo Alliance calls for 25 percent of electricity to be produced by renewables, a reduction of oil consumption by 25 percent, and improvements in high-performance buildings and regional transportation. It cites a significant number of cities like Albuquerque, which has already signed a City Renewable Energy Initiative. It would give tax incentives up to $1 million for solar manufacturers or solar R&D companies who move in, retrofit city-owned buildings with solar power, give rebates to residents who use renewables, and create new procurement standards for the city to lower greenhouse gas emissions, reduce toxic chemicals, and maximize energy and water efficiency.[24]

Pushing hard for renewable energy is important because of the way in which many influential establishment studies are interpreted. One of the best is the "wedge" theory developed by Robert Socolow and others at Princeton University, with backing from auto manufacturers like the Ford Motor Company. Socolow and his partners project global CO_2 emissions to grow from about 7 billion tons annually to about 15 billion tons by 2054. In order to reach stabilization, the United States and other nations would have to first reduce emissions and then hold them steady—that is, at a level that would appear as a flat line on a graph. The resulting triangles, or wedges, of cuts needed amount to about 7 billion tons. These are then divided up into seven possible wedges of 1 billion tons each. As Socolow puts it, no one solution or wedge can solve our needed emissions cuts alone. Some combination is needed. Socolow examined fifteen different "wedge" strategies, each of which "is based on known technology, is being implemented somewhere at an industrial scale, and has the potential to contribute a full wedge to carbon mitigation."[25] The fifteen strategies, such as forest plantation, fuel efficiency, and carbon sequestration and storage, are grouped into five categories: energy conservation, renewable energy, enhanced natural sinks, nuclear energy, and fossil carbon management. Clearly, relying only on available technologies that are already at industrial scale is a very conservative strategy. As is too often the case, such a cautious approach, designed to be persuasive, does not set high-enough achievable goals. Nevertheless, it then gets interpreted as the most "realistic" or "likely" scenario.

But looking at just cars, vans, SUVs, and light trucks as one of Socolow's wedges, we see that it is based on a study by the global auto industry's Sustainable Mobility Project (SMP). It reported 0.8 billion tons of CO_2 emissions as of 2000, one-eighth of all global emissions. The SMP predicts that light vehicle emissions will double by 2050 to 1.6 billion tons of carbon, based on increases of 600 million to 1.6 billion vehicles, averaging 10,000 miles per year at 30 miles per gallon. Socolow writes that nearly a whole wedge of reductions could be obtained by halving driving distance through more mass transit or telecommuting, *or* by doubling fuel economy.[26] But given that today's Toyota Prius already averages between 40 and 60 mpg, and cars with fuel economy of up to 100 mpg are about to be introduced, a tripling of fuel economy seems far more likely. Couple that with reductions in driving distance *as well,* and two wedges of the reduction pie seem quite achievable.

The same conservative assumptions hold true, for example, with energy efficiency in buildings and appliances. The IPCC estimated, for example, that two wedges of savings could be had through installing efficient lighting and available appliances, along with insulation. About half these savings may happen without incentives anyhow, according to the IPCC, so Socolow counts only one wedge—even though far greater efficiency than current technology offers is very likely to occur by 2054.[27] Similarly, low projections for cuts, savings, or wedges of CO_2 emissions are discussed for wind and solar power to replace coal. Again, the scenario considered is *either* a 50-fold expansion of wind *or* a 700-fold expansion of solar by 2054.[28] Other technologies like tidal and wave energy are simply discounted as untested. Socolow sees rapid growth in wind or solar energy as possible, but he also emphasizes possible constraints, such as the size and visibility of some sites and some opposition to current wind turbines (five stories high, with 300-foot rotors) and wind farms. But here, too, Socolow, in trying to be prudent, does not even count technologies that are not yet at "scale." Microgeneration of electricity at homes and small businesses that is now available using smaller wind turbines is not even discussed. Nevertheless, eighty thousand homes in Great Britain are already being powered by small, at-home power generation such as consumer-size wind turbines.[29] In 2006, David Cameron, leader of Britain's Conservative Party, announced he would add a wind turbine and solar panels to his west London home, giving microgeneration a huge boost. The British government has launched an initiative that will devote $150,000,000 over three years to encourage homeowners to move to microgeneration. Smaller home turbines are already being sold at retailers like B&Q, a chain of hardware stores, and sell for as little as $2,800. They can save about a third of a typical electric bill, with the government paying about 30 percent of the installed cost. By themselves, home-installed wind turbines and solar devices, already available, could supply about 8 percent of Britain's electricity and reduce carbon emissions by nearly 10 percent.[30]

Examining influential studies such as Socolow's is important, not because it is inaccurate, but because it reflects well the reduced expectations and assumptions that have shaped the conventional wisdom in the United States as a result of the environmentally unfriendly Bush years. Such careful and cautious arguments, despite their benign intent, can be misused by those who would promote nuclear power and coal. The nuclear component, for example, Socolow carefully explains, could save the 1 billion tons of CO_2 required for a full wedge only if an entirely new set of advanced design reactors were built, *tripling* current nuclear electricity capacity, and building seven hundred *new* 1,000-megawatt reactors to replace coal-fired ones. Socolow does review the considerable problems of waste and proliferation. And he adds that "nuclear power does *not* [emphasis added] have to play a major role in reducing global CO_2 emissions. . . . As providers of low-carbon electricity, wedges of nuclear power compete with wedges of wind and PV."[31] Nevertheless, in his conclusion, he states that one way to think about beginning to reduce emissions in the short run might be to implement 20 percent reductions from *each* of seven wedges. By indicating there may be short-term political advantage to including nuclear and coal energy sources, Socolow opens the door for industries and practices that he has already argued, rather convincingly, are not needed at all.

This is a critical point to understand when presenting shorter-term, initial plans for CO_2 reductions of interest to policymakers and candidates. A technical and cautious expert like Socolow does *not* say that we need coal and nukes to meet necessary carbon reductions. He is in fundamental agreement with environmental groups. Yet, too often, environmental and public health groups are painted as "advocates" with some sort of liberal bias. Nevertheless, their energy reports and studies lay out their assumptions, make clear what time lines and technologies they are talking about, and are quite credible. Unfortunately, they have been also labeled as "optimistic" by their critics, and as hopelessly compromised and inadequate by some grassroots environmentalists. They do not offer particularly radical proposals. They want to change policy, too. With a new Democratic, more environmentally friendly Congress, and then a new president, environmentalists will undoubtedly be bolder and are likely to be listened to more than at any time in the past decade. Their recommendations for the shorter term make excellent guides for talking to policymakers and candidates. Because major groups often strategize and lobby in concert, many of their proposals are similar. I have combined some of the main ones from the Sierra Club, U.S. Public Interest Research Group (U.S. PIRG), and Natural Resources Defense Council (NRDC). Their broad goals are spelled out in proposed legislation like the Boxer-Sanders Senate bill that calls for bold measures and an 80 percent cut in overall carbon emissions by 2050. But because it is not likely to pass while George W. Bush remains president, they also advocate initial steps that, at a minimum, would reduce American carbon emissions by 20 percent by the year 2020.[32]

COMMIT TO CAPS ON CARBON EMISSIONS

Mandatory caps on carbon emissions are necessary to solve climate change. Voluntary measures have been used since the United States agreed to the Rio treaty on climate in 1992—some fifteen years ago. They have not worked. The U.S. Senate should ratify the Kyoto Protocol and move on to an even stronger one promoted by the next administration. A strong cap-and-trade bill (without nuclear power) should be passed. This would put caps on power plants and other emitters and provide powerful incentives to minimize the cost of reducing pollution.

Increase Vehicle Fuel Economy Standards (CAFE) to at Least 40 mpg, and Include Light Trucks

The fuel economy of the fleet of new American cars, vans, and SUVs is worse now than it was during the Reagan administration. The creation of CAFE standards during the Carter administration and the oil embargoes and crises of the 1970s actually succeeded in reducing gas consumption and foreign oil imports. It also lowered corresponding CO_2 emissions and other pollutants. This 40 mpg figure can be reached within the next decade simply using technologies that already exist. Although 35 mpg CAFE standards were passed at the end of 2007, the time frame is too long, the fuel economy average too low. Beginning in 2009, the new 111th Congress should regularly review and raise CAFE standards as technologies improve.

Replace at Least 10 Percent of Gasoline in Cars with Biofuels and Other Clean Alternatives

As we saw in chapter 8, today's engines can use 10 percent ethanol right now, and higher percentages of biofuels (as much as 85 percent) can be used with available engine modifications for flexible fuel use. Congress should provide incentives for biofuels and reduce or eliminate tariffs on sugar-based fuels from Brazil and elsewhere. It should aggressively promote R&D in cellulosic ethanol and newer hybrid, electric, and fuel cell vehicles.

Require 20 Percent of Electricity from New Renewable Fuels

More than thirty states now have renewable portfolio standards (RPSs). The Congress should pass a federal RPS of at least 20 percent and sharply shift subsidies and R&D spending away from oil, coal, and gas, and toward renewable wind, solar, biomass, and other forms of clean, renewable energy. Failure to do so was one of the major disappointments of the

2007 energy bill. Personal commitments by members of Congress and administration officials to reduce emissions and install or purchase renewable power sources in their own homes should also be strongly recommended. The Congress and White House, followed by other federal buildings should be retrofitted for energy efficiency. Nearly $600 million has already been spent at the Capitol alone for security measures and an underground Visitors' Center whose plan lay dormant until the attacks on September 11.[33] A similar commitment to energy improvements should be made.

Energy Consumption in Homes, Businesses, and Industry Should Be Reduced by at Least 10 Percent

Measures to encourage improved energy efficiency should include tax breaks and credits for improving energy systems and efficiency, stronger regulatory standards for energy efficiency for appliances such as refrigerators, and greater R&D efforts and higher standards for green buildings and industry. Examples already exist, such as the Senate's Snowe-Feinstein bill and the corresponding Cunningham-Markey bill in the House. With the 111th Congress, even stronger measures should be possible.

These initiatives, and others like them, are readily achievable. They would significantly improve our air quality and health, reduce oil dependence, and cut CO_2 emissions by 20 percent by 2020. But newer and bolder measures will still be required. Citizens can and should press for greater change, while understanding that measures such as those outlined by environmental groups are best seen as an opening gambit, a minimum commitment that should be agreed to quickly by the Congress. In the 2008 election and long after, newer proposals, new legislation, new ideas will continue to emerge, especially if we continue to change the political climate. But politics is politics. There are still massive corporate forces and their allies aligned against the transformation of our outmoded, unhealthy, fossil fuel economy.

The carbon lobby and its political allies will continue to cast doubt on climate science and on the need for change, especially government action. They will continue to deploy squadrons of high-priced lobbyists and back candidates willing to delay and demagogue. Once again, a choice is put before us—between life and death, between participation and passivity.

Mr. and Mrs. Noah, there is a gathering storm. We risk the Flood. You and I are asked to do *something,* to build an ark, to save ourselves, and other species. We've got planks and nails and caulking and sails. Let's build a platform for others to walk on board. We *can* change the climate. That is the hope for a heated planet.

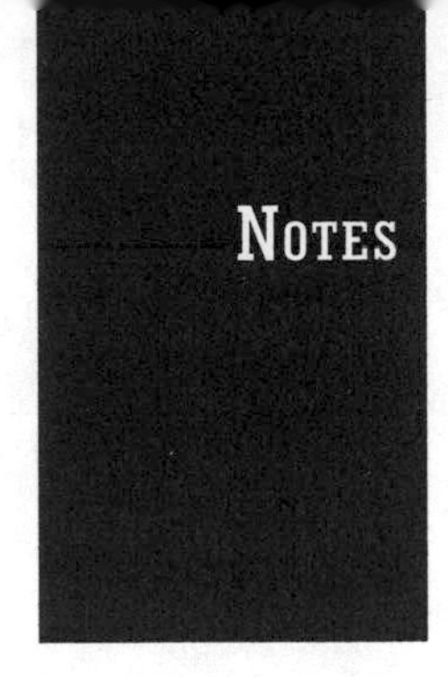

Notes

Introduction

1. Lawrence Wittner, *The Struggle against the Bomb,* vol. 2, *Resisting the Bomb: A History of the World Nuclear Disarmament Movement, 1954–1970* (Stanford: Calif., Stanford University Press, 1997), 416–423 and ix–x.

2. R. P. Turco, R. Toon, T. P. Ackerman, J. B. Pollack, and C. Sagan, "Global Atmospheric Consequences of Nuclear War," *Science* 222 (1983): 1,283. This original study was followed up by the same authors seven years later in "Climate and Smoke: An Appraisal of Nuclear Winter," *Science* 247 (1990): 166–176. Numerous studies since have corroborated the essential thesis. Sagan and Turco also published books on nuclear winter. Carl Sagan et al., *The Nuclear Winter: The World after Nuclear War* (London: Sidgwick and Jackson, 1985); and Carl Sagan and Richard Turco, *A Path Where No Man Thought: Nuclear Winter and the End of the Arms Race* (New York: Random House, 1990).

3. Wittner, *Resisting the Bomb,* 176–177, 195–197, 318–321, 371–373; and David Cortright, *Peace Works: The Citizen's Role in Ending the Cold War* (Boulder, Colo.: Westview Press, 1993).

4. After decades of work by groups like PSR, International Physicians for the Prevention of Nuclear War (IPPNW), and others, the first official government document calling for the complete abolition of nuclear weapons was published on August 14, 1996, by the Canberra Commission, initiated by the Australian government. Ron McCoy, co-president of IPPNW; Robert S. McNamara; Jacques Cousteau; and Gen. Lee Butler, former head of STRATCOM, were among the members. See http://www.dfat.gov.au/cc/cchome.html. In July 1996, the World Court also issued a historic opinion declaring nuclear weapons illegal. This, too, followed a lengthy, global grassroots campaign called the World Court Project. Complete information is on the Web site of the Lawyers' Committee for Nuclear Policy at http://www.lcnp.org/wcourt. Widespread campaigns and protests worldwide and lobbying in the United States also led to the end of nuclear testing and the signature of the Comprehensive Nuclear-Test-Ban Treaty (CTBT) by President Clinton in September 1996.

5. Jim Hansen, "The Threat to the Planet," *New York Review of Books* 53, no. 12, July 13, 2006, at http://www.nybooks.com/articles/19131. Accessed on July 17, 2006. Hansen uses high-end estimates to indicate that sea-level rise of up to 80 feet is possible with worst-case scenarios. Also see Marc Kaufman, "Escalating Ice Loss in Antarctica," *Washington Post,* January 14, 2008.

6. Michael Shellenberger and Ted Nordhaus, "The Death of Environmentalism: Global Warming Politics in a Post-Environmental World" (Breakthrough Institute: September 2004), at http://thebreaktrough.org/images/Death_of_Environmentalism.pdf. Also see Michael Shellenberger and Ted Nordhaus, *Break Through: From the Death of Environmentalism to the Politics of Possibility* (New York: Houghton Mifflin, 2007).

Chapter 1. Understanding Climate Change

1. Andrew C. Revkin, "Big Arctic Perils Seen in Warming," *New York Times,* October 30, 2004. For an in-depth report on Arctic warming, see ACIA, *Impacts of a Warming Arctic: Arctic Climate*

Impact Assessment (Cambridge: Cambridge University Press, 2004) and at http://www.acia.uaf.edu/pages/overview.html.

2. For an analysis and graphs of how global average annual temperature is determined, see the NASA Goddard Space Institute Web site at http://data.giss.nasa.gov/gistemp/2005. The year 2005 was technically in a statistical tie with 1998. But because 1998 was an El Niño–affected year, the average for 2005 is even more ominous. In 2007, a statistically insignificant error in data adjustment was found in NASA global averages. U.S. records from NOAA were not involved. Only Rush Limbaugh bothered to complain. See Marc Kaufman, "NASA Revisions Create a Stir in the Blogosphere," *Washington Post,* August 15, 2007. As average temperatures for 2007 were being calculated in early 2008, another record year was about to be posted. See http://www.sciencedaily.com/releases/2007/01/070105080024.htm.

3. The *Fourth IPCC Assessment Report,* by Working Group I, on the scientific basis of global warming is available from the National Center for Atmospheric Research at http://www.ipcc.ch/ipccreports/ar4-wg1.htm.

4. For a version of the classical oath used in Greek times, see http://www.pbs.org/wgbh/nova/doctors/oath_classical.html. Modern versions of the oath have deleted references to the Greek gods, abortion, and other issues and often do not contain the admonition to do no harm. But the culture of prevention and taking action, of equal treatment for all, and of caring about injustice remains strong.

5. Dade W. Moelle, *Environmental Health* (Cambridge, Mass.: Harvard University Press, 1992), 4–8. For an excellent overview of the public health approach to the environment, see Michael McCally, "Environment, Health, and Risk," chap. 1 in Michael McCally, ed., *Life Support: The Environment and Human Health* (Cambridge, Mass.: MIT Press, 2002).

6. For Snow's contributions to epidemiology, see George Rosen, *A History of Public Health,* expanded ed. (Baltimore: Johns Hopkins University Press, 1993), 261–264. For the full Snow story and the implications of his work for contemporary times, see Steven Johnson, *The Ghost Map: The Story of London's Most Terrifying Epidemic—and How It Changed Science, Cities, and the Modern World* (New York: Riverhead Books, 2006). A comprehensive biography with original documents is Peter Vinten-Johansen et al., *Cholera, Chloroform, and the Science of Medicine: A Life of John Snow* (Oxford: Oxford University Press, 2003).

7. David Patrick Houghton, *U.S. Foreign Policy and the Iran Hostage Crisis* (Cambridge: Cambridge University Press, 2001), 2–4.

8. Intergovernmental Panel, *Climate Change 2001: The Scientific Basis. Contribution of Working Group I to the Third Assessment Report of the IPCC,* ed. J. T. Houghton et al. (Cambridge: Cambridge University Press, 2001); online at http://www.ipcc.ch/pub/reports.htm.

9. Spencer R. Weart, *The Discovery of Global Warming* (Cambridge, Mass.: Harvard University Press, 2003), 191, 196.

10. Elisabeth Rosenthal and Andrew S. Revkin, "Science Panel Calls Global Warming 'Unequivocal,' " *New York Times,* February 3, 2007. The mention of surprises can be found in *Summary for Policy Makers: The Science of Climate Change—IPCC Working Group I,* section 6, "There Are Still Many Uncertainties," at http://www.ipcc.ch/pub/sarsum1.htm.

11. Ross Gelbspan, *The Heat Is On: The Climate Crisis, the Cover Up, the Prescription,* updated ed. (New York: Perseus Books, 1998), 1–2. First published in 1997.

12. Ross Gelbspan, *Boiling Point: How Politicians, Big Oil and Coal, Journalists, and Activists Have Fueled the Climate Crisis—and What We Can Do to Avert Disaster* (New York: Perseus Books, 2004), 8–9, 19–21.

13. Robert K. Musil, "The Political and Public Health Implications of Global Warming," *P&S Medical Review* 8, no. 1 (Fall 2001): 3–7. Available at http://www.cumc.columbia.edu/news/review/pdf/f3_Musil.pdf. Accessed August 14, 2007. Original reports indicated thirty-five thousand deaths in Europe from the 2003 heat wave. Later, more conclusive data indicted fifty-two thousand deaths. See Janet Larsen, "Setting the Record Straight: More than 52,000 Died from Heat in Summer 2003," Earth Policy Institute update, July 28, 2006. Available at http://www.earth-policy.org/Updates/2006/Update56.htm. Accessed August 14, 2007.

14. Juliet Eilperin, "Warming Tied to Extinction of Frog Species," *Washington Post*, January 12, 2006.

15. Blaine Harden, "Experts Predict Polar Bear Decline: Global Warming Is Melting Their Ice Pack Habitat," *Washington Post*, July 7, 2005.

16. Earth Policy Institute, "U.S. Leads World in Climate Refugees," August 23, 2006, at http://www.enn.com/net_PF.html?id=1621. Accessed August 28, 2006. For Katrina casualty estimates and summaries of other data, see "Hurricane Katrina" in Wikipedia at http://en.wikipedia.org/wiki/Hurricane_Katrina. Accessed September 11, 2006.

17. Weart, *Discovery of Global Warming*, 21–24.

18. Ibid., 35–38.

19. Paul Andrew Mayewski and Frank White, *The Ice Chronicles: The Quest to Understand Global Climate Change* (Hanover, N.H.: University of New Hampshire/University Press of New England, 2002). Also, Al Gore, *An Inconvenient Truth: The Planetary Emergency of Global Warming and What We Can Do About It* (Emmaus, Pa.: Rodale Press, 2006), 66–67.

20. Mark Maslin, *Global Warming: A Very Short Introduction* (Oxford: Oxford University Press, 2004), 6.

21. Remarks by Dan Lashof, American University, March 2006, and in an unpublished memo by Dan Lashof, Natural Resources Defense Council, "Global Warming Science Update: Year End Review—2005." Also see C. A. Mears and F. J. Wentz, "Amplification of Surface Temperature Trends and Variability in the Tropical Atmosphere," *Science* 309 (August 11, 2005, online): 1,548; and Andrew Revkin, "Errors Cited in Assessing Climate Data," *New York Times*, August 12, 2005.

22. D. Lashof, remarks at American University; J. Hansen et al., "Earth's Energy Imbalance: Confirmation and Implications," *Science* 308 (April 28, 2005, online): 1,431; and T. P. Barnett et al., "Penetration of Human-Induced Warming into the World's Oceans," *Science* 309 (June 2, 2005, online): 284.

23. Weaver, a Canadian scientist, comments in " 'Smoking Gun' Report to Say Global Warming Here," January 23, 2007, CNN.com at http://www.cnn.com/2007/TECH/science/01/23/climate.report.ap/index.html?section=cnn_latest. Accessed February 9, 2007.

24. The "hockey stick" graph of CO_2 and temperature correlations had been challenged on obviously political grounds by Representative Joe Barton (R-TX), when he was chairman of the House Energy and Commerce Committee. Using analyses from climate skeptics, Barton claimed that the 1998 work of Michael E. Mann, Ray S. Bradley, and Malcolm K. Hughes that forms the basis of much of the hockey stick had statistical errors and used some flawed data. Barton's charges were objected to by the president of the National Academy of Sciences, the executive director of the American Association for the Advancement of Science, and Barton's Republican colleague, Sherwood Boehlert (R-NY), a leading and knowledgeable environmental champion. Analyses by scientists at the National Center for Atmospheric Research have validated Mann and his colleagues while rejecting totally the claims of their critics (Lashof, American University lecture and fact sheet). Malcolm K. Hughes, professor of dendrochronology at the University of Arizona, offers a detailed response in a July 15, 2005, letter to Chairman Barton at http://www.realclimate.org/Hughes_response_to_Barton.pdf. Also see E. R. Wahl and C. M. Ammann, "Robustness of the Mann, Bradley, Hughes Reconstruction of Surface Temperatures: Examination of Criticisms Based on the Nature and Processing of Proxy Climate Evidence," *Climatic Change* (forthcoming).

25. For an explanation of the increasingly extreme hydrological cycle, see John Houghton, *Global Warming: The Complete Briefing*, 3rd ed. (Cambridge: Cambridge University Press, 2004), 128–131.

26. Gelbspan, *Boiling Point*, 21–22.

27. Brenda Fowler, "Scientists Enthralled by Bronze Age Body," *New York Times*, October 1, 1991. A full account of the publicity surrounding this find as a result of glacial melt is in Brenda Fowler, *Iceman: Uncovering the Life and Times of a Prehistoric Man Found in an Alpine Glacier* (Chicago: University of Chicago Press, 2001).

28. Photos of retreating glaciers and information are at http://www.worldviewofglobal warming.org/pages/glaciers.html. Also see Gore, *An Inconvenient Truth,* 42–59; and various studies by Lonnie Thompson, who is profiled in *National Geographic*, "The Vanishing World of Lonnie Thompson, August 2004, at http://www.nationalgeographic.com/adventure/0408/excerpt5.html.

29. James Hansen, "The Threat to the Planet," *New York Review of Books*, 53, No. 12, July 13, 2006, at www.nybooks/articles/19131. Accessed July 17, 2006. Studies in 2007 indicated that ice was melting even faster than models have predicted. See Nicholas Kristof, "The Big Melt," *New York Times*, August 16, 2007; and Andrew C. Revkin, "Analysts See 'Simply Incredible' Shrinking of Floating Ice in the Arctic," *New York Times*, August 10, 2007.

30. *Summary for Policy Makers: The Science of Climate Change—IPCC Working Group I*, section 6, "There Are Still Many Uncertainties." The science section of the final version of the Fourth Assessment (AR4) is at http://www.ipcc.ch/ipccreports/ar4-wg1.htm. The full Final Report of the Fourth Assessment, released November 17, 2007, is available at http://www.ipcc.ch. For the best review of climate surprises, see the executive summary from Committee on Abrupt Climate Change, National Academy of Sciences, *Abrupt Climate Change: Inevitable Surprises* (Washington, D.C.: National Academies Press, 2002), at http://books.nap.edu/execsumm_pdf/10136.pdf.

31. John D. Cox, *Climate Crash: Abrupt Climate Change and What It Means for Our Future* (Washington, D.C.: John Henry Press, 2005), 164–167.

32. Former Secretary of State Colin Powell's full remarks to the WSSD at the Sandton Convention Center, September 4, 2004 are at http://www.state.gov/secretary/former/powell/remarks/2002/13235.htm. Accessed July 30, 2006. For typical press coverage and a photo of protesters being removed from the hall, see James Dao, "Protesters Interrupt Powell Speech as UN Talks End," *New York Times,* September 5, 2002.

33. For a current review of children's health, see the Web site of the Global Health Council at http://www.globalhealth.org/view_top.php. Accessed July 30, 2006.

34. A. J. McMichael, *Planetary Overload: Global Environmental Change and the Health of the Human Species* (Cambridge: Cambridge University Press, 1995), 146, 151.

35. Tony McMichael, *Human Frontiers, Environments and Disease: Past Patterns, Uncertain Futures* (Cambridge: Cambridge University Press, 2001), 1–6.

36. *The Third Assessment Report of Working Group I: Summary for Policy Makers*, 14, at http://www.ipcc.ch/pub/spm22-01.pdf. The Fourth Assessment estimated slightly lower ranges, though not significant for the flooding of coastal areas. See "Summary for Policy Makers," section 5, at http://www.ipcc.ch/pub/sarsum1.htm#five.

37. Hansen, "The Threat to the Planet."

38. Paul R. Epstein and Evan Mills, eds., *Climate Change Futures: Health, Ecological and Economic Dimensions* (Cambridge, Mass.: Center for Health and the Global Environment, Harvard Medical School, 2005), 32. The Center's Web site is an excellent source of articles and reports on climate change and health at http://www.chge.med.harvard.edu.

39. http://www.malaria.org.zw/ecology.html#temp.

40. World Health Organization, *Report of the Commission on Macroeconomics and Health* (Geneva: World Health Organization, 2001). For a complete review of malaria and health studies and implications, see Physicians for Social Responsibility, *The Modern Malaria Handbook*, introduction by Robert K. Musil (Washington D.C.: PSR, 1998). The review of malaria and climate by Epstein and colleagues is in Epstein and Mills, *Climate Change Futures*, 32–35. Also see A. J. McMichael, A. Haines, R. Sloof, and S. Kovats, eds., *Climate Change and Human Health: An Assessment Prepared by a Task Group on Behalf of the World Health Organization, the World Meteorological Organization and the United Nations Environment Programme*, (Geneva: World Health Organization, 1996), 78–84.

41. Paul R. Epstein, "Is Global Warming Harmful to Health?" *Scientific American,* August 2000, 50–57.

42. Epstein and Mills, *Climate Change Futures*, 33.

43. Tarekegn Abeku et al., "Spatial and Temporal Variations of Malaria Epidemic Risk in Ethiopia: Factors Involved and Implications," *Acta Topica* 87, no. 3 (August 2003): 331–340.

44. Jonathan A. Patz et al., "Impact of Regional Climate Change on Human Health," *Nature* 438, no. 17 (November 2005): 311.

45. John Houghton, *Global Warming: The Complete Briefing*, 2nd ed. (Cambridge: Cambridge University Press, 1997), 132. In the 3rd ed., 178, Houghton does not offer an estimate of growth but simply says that malaria and dengue areas will increase, with more populations at risk.

46. Epstein and Mills, *Climate Change Futures*, 35.

47. Ibid., 34.

48. Ibid., 36.

49. Rita R. Colwell, "Global Climate and Infectious Disease: The Cholera Paradigm," *Science*, n.s., 274, no. 5295 (December 20, 1996): 2,025–2,031.

50. Patz et al., "Impact of Regional Climate Change on Human Health," 311.

51. Ibid.

52. Ibid., 313.

53. Devra Lee Davis et al., "Short-Term Improvements in Public Health from Global-Climate Policies on Fossil Fuel Combustion; an Interim Report," *Lancet* 350 (1997): 1,341–1,349.

54. Kim Knowlton et al., "Assessing Ozone-Related Health Impacts under a Changing Climate," *Environmental Health Perspectives* 112, no. 15 (November 2004): 1,557–1,563.

CHAPTER 2. HOME, HOME ON THE RANGE

1. For a contemporary account of the negotiations, see Robert K. Musil, "Global Climate Change: The View from Kyoto," *PSR Reports* 19, no. 1 (Winter 1998): 1, 6. Available at http://www.psr.org/documents/1998_01.pdf. For a key health study presented at Kyoto, see Devra Lee Davis et al., "Short-Term Improvements in Public Health from Global-Climate Policies on Fossil Fuel Combustion; an Interim Report," *Lancet* 350 (1997): 1,341–1,349. The official texts in various languages and other information are at http://unfccc.int/kyoto_protocol/items/2830.php. For useful summaries and data on Kyoto, also see http://en.wikipedia.org/wiki/Kyoto_Protocol.

2. *Death by Degrees* (Degrees of Danger) reports can be downloaded at the PSR Web site at http://www.psr.org/site/PageServer?pagename=enviro_resources#GlobalWarming.

3. Governor O'Malley's executive order is at http://www.gov.state.md.us/executiveorders/01.07.07ClimateChange.pdf. His announcement the action plan is at: http://www.gov.state.md.us/pressreleases/070420.html.

4. Jennifer Lenhart, "Public to Glimpse 'Uncle Tom's Cabin,' " *Washington Post*, June 8, 2006.

5. Maryland information is from the EPA's former global warming Web site at www.epa.gov/globalwarming/impacts/stateimp/maryland/index.html. Similar information from the EPA in 1998 can be found at http://yosemite.epa.gov/OAR/globalwarming.nsf/UniqueKeyLookup/SHSU5BUSTE/$File/md_impct.pdf. The Chesapeake Bay Foundation has done a more current report on the effects of warming on the ecology of the bay itself. See David A. Fahrenthold, "Warming Poses Threats to Chesapeake, Group Says," *Washington Post*, July 20, 2007. The full report, "Climate Change and the Chesapeake Bay: Challenges, Impacts, and the Multiple Benefits of Agricultural Conservation," is on the CBF Web site at http://www.cbf.org/site/DocServer/climatechange.pdf?docID=9423.

6. Physicians for Social Responsibility and Ozone Action, *Heat Waves and Hot Nights*, July 26, 2000. PSR and author's files.

7. The National Center for Health Statistics Web site where this information was accessed in 2001 is no longer available. For the most recent statistics, which show heat still the highest cause of fatalities by far, with an average of 150 per year over the past ten years, see the National Weather Service site on fatalities at http://www.nws.noaa.gov/om/hazstats.shtml.

8. Jan Semenza et al., "Heat-Related Deaths during the July 1995 Heat Wave in Chicago," *New England Journal of Medicine* 335 (1996): 84–90.

9. L. S. Kalkstein and J.S. Greene, "An Evaluation of Climate/Mortality Relationships in Large

U.S. Cities and the Possible Impacts of Climate Change," *Environmental Health Perspectives* 105 (1997): 84–93.

10. Elizabeth Thompson, *Poisoned Power: How America's Outdated Electric Plants Harm Our Health and Environment* (New York: Clean Air Network, 1997).

11. See the EPA greenhouse gas data from 2000 onward, plus projections, at http://epa.gov/climatechange/emissions/downloads/ch5.pdf.

12. U.S. PIRG, "Lethal Legacy: The Dirty Truth about the Nation's Most Polluting Power Plants," (April 2000), http://static.uspirg.org/reports/lethallegacy2000/LethalLegacy.PDF.

13. Thompson, *Poisoned Power*. The more recent figures are found in the Environment Maryland Research and Policy Center report, *Particulate Matter Pollution from Maryland Power Plants*, available from U.S. PIRG at http://www.uspirg.org/uploads/nI/oh/nIoh1HXd6iGM6fi9NOxLZA/Maryland-PM-Pollution.June2007.pdf. It should be noted that with tighter controls, NOx emissions are about 50 percent of what they were a decade ago. Additional compiled figures on pollution from Maryland Power Plants can be found in a 2003 report by Brendon Wu of the U.S. Public Interest Research Group Education Fund at http://cta.policy.net/reports/lethal_legacy_report.pdf.

14. See the EPA's "Fact Sheet on Health and Environmental Effects of Ground-Level Ozone," at http://www.epa.gov/ttn/oarpg/naaqsfin/o3health.html.

15. C. P. Weisel et al., "Relationship between Summertime Ambient Ozone Levels and Emergency Department Visits for Asthma in Central New Jersey," *Environmental Health Perspectives* 103, supp. 2 (1995): 97–102.

16. R. Delfino et al., "Effects of Air Pollution on Emergency Room Visits for Respiratory Illness in Montreal, Quebec," *American Journal of Respiratory Critical Care Medicine* 155 (1997): 568–576. For a thorough review of air pollution and health, see Howard Frumkin, Lawrence-Frank, and Richard Jackson, *Urban Sprawl and Public Health: Designing, Planning, and Building for Healthy Communities* (Washington, D.C.: Island Press, 2004), 79–89.

17. EPA, "Latest Findings on National Air Quality: 1999 Status and Trends." Web site no longer available.

18. U.S. PIRG, Clean Air Task Force, *Adverse Health Effects Associated with Ozone in the Eastern United States, 2000*, at http://www.pirg.org/reports/enviro/nobreath/report.pdf.

19. U.S. PIRG, Clean Air Task Force, *Death, Disease and Dirty Power: Mortality and Health Damage Due to Air Pollution from Power Plants* (October 2000). Available at http://www.net.org/relatives/4211.pdf.

20. For an excellent summary by the Environment News Service, see http://www.ens-newswire.com/ens/feb2006/2006–02–15–02.asp. The announcement of the report by the Maryland Nurses Association is at http://www.thehastingsgroup.com/marylandrn/press_release.html.

21. From the former EPA global warming Web site on sea-level rise at http://www.epa.gov/global warming/publications/impacts/sealevel/ocean_city.html. The most recent site, which is not as detailed, is at http://www.epa.gov/climatechange/effects/coastal/index.html#sea.

22. United States Geological Survey (USGS), "The Changing Bay: Geologic Product of Rising Sea Level," was at http://pubs.usgs.gov/factsheets/fs/102–98. No longer accessible.

23. EPA, "Sea-Level Rise at Ocean City." No longer available.

24. The EPA former fact sheet (only) is available at http://yosemite.epa.gov/OAR/global warming.nsf/UniqueKeyLookup/SHSU5BUSTE/$File/md_impct.pdf.

25. EPA, "Sea Level Rise at Ocean City."

26. EPA (James G. Titus and Charlie Richman), "Maps of Lands Vulnerable to Sea-Level Rise: Modeled Elevations along the U.S. Atlantic and Gulf Coasts." Formerly at http://www.epa.gov/globalwarming/publications/impacts/sealevel/maps/index.html. A longer and more recent technical analysis by Titus and Richman is at http://www.epa.gov/climatechange/effects/downloads/maps.pdf.

27. For a comprehensive look at efforts to restore oysters in the bay, see Carl F. Cerco and Mark R. Noel, "Evaluating Ecosystem Efforts of Oyster Restoration in the Chesapeake Bay: A Re-

port to the Maryland Department of Natural Resources" (U.S. Army Corps of Engineers Engineer Research and Development Center, 2005), at http://www.chesapeakebay.net/pubs/Cerco_Noel_final.pdf. The direct connection between oysters and salinity was documented at the former EPA global warming site, no longer available.

28. Howard Frumkin, MD, lecture at the Johns Hopkins School of Public Health, April 2001. Also see Howard Frumkin, Lawrence Frank, and Richard Jackson, *Urban Sprawl and Public Health: Designing, Planning, and Building for Healthy Communities* (Washington, D.C.: Island Press, 2004), 174, for differences in sense of community between Greenbelt and Hyattsville, Maryland, based on degrees of sprawl.

29. See the Chesapeake Bay Foundation for Land Use and the Chesapeake Bay at http://www.cbf.org/site/PageServer?pagename=resources_facts_sprawl. Data on vehicle miles traveled have been removed.

30. Joan Rose et al., "Climate and Waterborne Disease Outbreaks," *Pathogens* 92, no. 9 (September 2000): 78–87.

31. Centers for Disease Control, "Flood-Related Mortality—Georgia, July 4–14, 1994," *Morbidity and Mortality Weekly Report* 43 (July 29, 1994): 526–530.

32. U.S. PIRG, *Flirting with Disaster: Global Warming and the Rising Costs of Extreme Weather*, April 17, 2001, was at http://www.pirg.org/disaster/report.pdf. Contact U.S. PIRG staff for archival reports.

33. A. J. McMichael, A. Haines, R. Sloof, and S. Kovats, eds., *Climate Change and Human Health* (Geneva: World Health Organization, 1996).

34. Rose et al., "Climate and Waterborne Disease Outbreaks."

35. T. K. Graczyk et al., "Environmental and Geographical Factors Contributing to Watershed Contamination with *Cryptosporidium parvum* Oocysts," *Environmental Research* 82, no. 3 (March 2000): 263–271, at http://www.ncbi.nlm.nih.gov/sites/entrez?cmd=Retrieve&db=PubMed&list_uids=10702335&dopt=Abstract.

36. See Rodney Barker, *And the Waters Turned to Blood* (New York: Simon and Schuster, 1997), for a full account of *Pfiesteria*.

37. Author's interview with David Brubaker, GRACE, April 2001. See www.factoryfarm.org for information on GRACE, the Eastern Shore, and other areas. Also see the Johns Hopkins School of Public Health project, Center for a Livable Future, at http://www.jhsph.edu/clf.

38. A DVD can be ordered from the Chesapeake Climate Action Network at http://chesapeakeclimate.org/pages/page.cfm?page_id=164.

39. Author's telephone interview with Rep. Dan K. Morhaim, MD, Baltimore, April 27, 2001. Morhaim has a Web page at http://www.drdanmorhaim.com. It also links to Maryland legislative information at http://www.mlis.state.med.us.

40. Morhaim et al., House Bill 276, Fiscal Note, Maryland General Assembly, 2001 session. http://www.mlis.state.md.us/2001rs/billfile/HB0276.htm.

41. Jim Motavalli with Sherry Barnes, "Greater New York: Urban Anxiety," in Jim Motavalli, ed., *Feeling the Heat: Dispatches from the Frontlines of Climate Change* (New York: Routledge, 2004), 40–41.

42. Ibid., 42–44. Also see Janine Bloomfield, "Hot Nights in the City: Global Warming, Sea-Level Rise and the New York Metropolitan Region," Environmental Defense, 1999, at http://www.environmentaldefense.org/documents/493_HotNY.pdf.

43. Paul R. Epstein and Evan Mills, eds., *Climate Change Futures: Health, Ecological and Economic Dimensions* (Cambridge, Mass.: Center for Health and the Global Environment, Harvard Medical School, 2005), 41–43. Figures for 2005 and 2006 are from the Centers for Disease Control at http://www.cdc.gov/ncidod/dvbid/westnile/surv&controlCaseCount06_detailed.htm.

44. Al Gore, *An Inconvenient Truth: The Planetary Emergency of Global Warming and What We Can Do about It* (Emmaus, Pa.: Rodale Press, 2006), 208–209.

45. Physicians for Social Responsibility, *Death by Degrees: The Emerging Health Crisis of Climate Change in New Hampshire* (November 1999), at http://www.psr.org/site/DocServer/

Death_By_Degrees_New_Hampshire.pdf?docID=550. Also see the front-page report on the release of PSR's first state report in New Hampshire in *PSR Reports* at http://www.psr.org/documents/2000_01.pdf.

46. PSR, *Death by Degrees: . . . New Hampshire*.

47. Ibid.

48. Frumhoff's comment is to Reuters. See Timothy Gardner, "US Northeast Faces Flood Risks from Global Warming," *Planet Ark News*, July 13, 2007, at http://www.planetark.com/avantgo/dailynewsstory.cfm?newsid=43076. Accessed July, 13, 2007. The full UCS report, *Confronting Climate Change in the U.S. Northeast* (July 2007), is available at http://www.climatechoices.org/assets/documents/climatechoices/confronting-climate-change-in-the-u-s-northeast.pdf. Accessed July 23, 2007.

49. A report on one of the air-quality studies by PSR on the border, "A Retrospective Study on Pediatric Asthma and Air Quality," is at http://yosemite1.epa.gov/oia/MexUSA.nsf/641f7d7486c7e19b882565a8005f769d/746025277a103876882567c500623662!OpenDocument. Accessed July 23, 2007. Another brief description of a grant to the PSR Project and other environmental justice grants is at http://www.purdue.edu/dp/envirosoft/grants/src/ej/regg6.htm.

50. Physicians for Social Responsibility, *Degrees of Danger: How Smarter Energy Choices Can Protect the Health of Texans* (June 2003), 5, 12–13, 16. Available at www.psr.org/site/DocServer/Degrees_of_Danger_Texas.pdf?docID=557.

51. Ibid., 6–7.

52. Ibid. For more information on the toxic effects of mercury and the campaign to halt them, see the PSR Web site for the campaign led by Susan West Marmagas, MPH, and Karen Perry, MPA, at http://mercuryaction.org.

53. PSR, *Degrees of Danger: How Smarter Energy Choices Can Protect the Health of Texans*, 7–8, 23–26.

54. PSR, *Degrees of Danger: The Emerging Health Crisis of Climate Change in Georgia* (February 2000), 13–14. Available at www.psr.org.

55. Terry Tamminen, *Lives per Gallon: The True Cost of Our Oil Addiction* (Washington, D.C.: Island Press, 2006).

CHAPTER 3. THE POWER OF THE CARBON LOBBY

1. For an overview of corporate activities to cover up the dangers of lead, especially as an additive in gasoline, see Jamie Lincoln Kitman, "The Secret History of Lead," *Nation*, March 20, 2000. For a quick history of tobacco industry propaganda, see John C. Stauber and Sheldon Rampton, *Toxic Sludge Is Good for You: Lies, Damn Lies and the Public Relations Industry* (Compton, Maine: Common Courage Press, 1995), 25–32. Also see Larry C. White, *Merchants of Death: The American Tobacco Industry* (New York: William Morrow, 1988).

2. Michael T. Klare, *Blood and Oil: The Dangers and Consequences of America's Growing Dependency on Imported Petroleum* (New York: Metropolitan Books, 2004), 36.

3. Ibid., 30.

4. Ibid., 43.

5. Carter quoted in ibid., 46. For Carter's full speech, see "Jimmy Carter: State of the Union Address: January 23, 1980," at http://www.jimmycarterlibrary.org/documents/speeches/su80jec.phtml.

6. Bush and Cheney quoted in Klare, *Blood and Oil*, 50.

7. Cheney quoted in Klare, *Blood and Oil*, 99. See also the transcript of Cheney's remarks in *New York Times*, August 26, 2002.

8. Jeremy Leggett, *The Carbon War: Global Warming and the End of the Carbon Era* (New York: Routledge, 2001), 5–8.

9. Ibid., 10–11.

10. Ross Gelbspan, *Boiling Point: How Politicians, Big Oil and Coal, Journalists, and Activists Have Fueled the Climate Crisis—and What We Can Do to Avert Disaster* (New York: Basic Books), 43–44.

11. Luntz quoted in ibid., 41.

12. Gelbspan, *Boiling Point*, 45–46. For Peabody operations worldwide, including the sale in 2006 of 248 million tons of coal and revenues of $5.3 billion, see http://www.peabodyenergy.com.

13. Gelbspan, *Boiling Point*, 47–50.

14. Ibid., 54–56.

15. Ibid., 56–58.

16. U.S. House of Representatives, Committee on Oversight and Government Reform, letter to James Connaughton, Chair, Council on Environmental Quality, January 22, 2007. Video and transcripts of the March 19, 2007, Waxman hearings from the House Committee on Oversight and Government Reform are at http://oversight.house.gov/story.asp?ID=1214.

17. Jeff Goodell, *Big Coal: The Dirty Secret behind America's Energy Future* (New York: Houghton Mifflin, 2006), 131. For a full analysis of the environmental and public health campaign over NAAQS, see "Clinton Strengthens Clean Air Standards," *PSR Reports* 18, no. 2 (Summer 1997), at http://www.psr.org/documents/1997_03.pdf.

18. Goodell, *Big Coal*, 131, including the quotation about Schwartz.

19. See Carol Browner's testimony on the new standards at http://www.resourcescommittee.house.gov/archives/105Cong/fullcomm/Sep30.97/browner.htm. For more information on the NAAQs, see http://www.epa.gov/ttn/oarpg/naaqsfin/naaqs.htm.

20. Various fact sheets and information on mercury are available from PSR at http://www.psr.org/site/DocServer/Mercury_Fact_Sheet__1.pdf?docID=708. For a full review of mercury and other pollutants, see Clear the Air/PSR, *Children at Risk: How Air Pollution from Power Plants Threatens the Health of America's Children* (April 2002), available at http://www.psr.org/site/DocServer/ChildrenatRisk.pdf?docID=529.

21. Shea quoted in Goodell, *Big Coal*, 138.

22. Michael Shore, *Out of Control and Close to Home: Mercury Pollution from Power Plants*, Environmental Defense, 2003, at http://www.environmentaldefense.org/documents/3370_MercuryPowerPlants.pdf. Also see Juliet Eilperin, "Mercury 'Hot Spots' Identified in U.S. and Canada," *Washington Post*, January 6, 2007.

23. Randy Lee Loftis, "Texas Cool to Confront Global Warming," *Dallas Morning News*, September 3, 2006. Accessed at http://www.dallasnews.com/cgi-bin/bi/gold_print.cgi on September 6, 2006.

24. Ibid.

25. Ibid.

26. Roddy Scheer, "Greens Step Up Pressure to Stop New Texas Coal Plants," *E: The Environmental Magazine*, December 18, 2006, at http://www.emagazine.com/view/?3514&printview&imagesoff. Protests included hunger strikes. See Eric Griffey, "Taking Lumps over Coal: Plans for 17 New Coal-Fired Plants Are Drawing Howls from All Over the Map," *FW Weekly*, December 13, 2006, at http://www.fwweekly.com/content.asp?article=4452. See also Ralph Blumenthal, "Texans Beat Big Coal, and a Film Shows How," *New York Times*, April 5, 2008. For the film, *Fighting Goliath: Texas Coal Wars*, narrated by Robert Redford, see http://www.fightinggoliathfilm.com.

27. Andrew Ross Sorkin, "A $45 Billion Buyout Deal That Has Many Shades of Green," *New York Times*, February 26, 2007. Also see Elizabeth Souder, "TXU Bidders Would Cut 8 of 11 Proposed Plants, *Dallas Morning News*, February 25, 2007. Comments by Environmental Defense's Fred Krupp, who helped broker the deal, are at http://www.environmentaldefense.org/article.cfm?contentID=5983.

28. Goodell, *Big Coal*, 43.

29. Elizabeth Williamson, "The Green Gripe with Obama: Liquefied Coal Is Still . . . Coal," *Washington Post*, January 10, 2007.

30. Goodell, *Big Coal*, 62–63, including the quotation of Lauriski.

31. Ibid., 78.

32. Gelbspan, *Boiling Point*, 43.

CHAPTER 4. FRAMING AND TALKING ABOUT GLOBAL WARMING

1. Michael Shellenberger and Ted Nordhaus, "The Death of Environmentalism," available at http://www.thebreakthrough.org/images/Death_of_Environmentalism.pdf.

2. Clear the Air, *Dirty Air, Dirty Power: Mortality and Health Damage Due to Air Pollution from Power Plants* (June 2004), available at http://www.cleartheair.org/dirtypower/docs/dirtyAir.pdf.

3. See, for example, Physicians for Social Responsibility, *Emerging Links: Chronic Disease and Environmental Exposure—Parkinson's Disease* (2003), available at http://www.psr.org/site/DocServer/Emerging_Links_Parkinson_s_Disease.pdf?docID=702; or PSR, *Bearing the Burden: Health Implications of Environmental Pollutants in Our Bodies* (January 2003), available at http://www.psr.org/site/DocServer/Bearing_the_Burden_summary.pdf?docID=1361.

4. George Lakoff, *Moral Politics: How Liberals and Conservatives Think* (Chicago: University of Chicago Press, 2002); and George Lakoff, *Don't Think of an Elephant: Know Your Values and Frame the Debate*, foreword by Howard Dean, introduction by Don Hazen (White River Junction, Vt.: Chelsea Green, 2004).

5. Lakoff, *Moral Politics*, 422–423.

6. Andrew C. Revkin, "Big Arctic Perils Seen in Warming," *New York Times*, October 30, 2004. For an in-depth report on Arctic warming, see ACIA, *Impacts of a Warming Arctic: Arctic Climate Impact Assessment* (Cambridge: Cambridge University Press, 2004), and at www.acia.uaf.edu/pages/overview.html.

7. See the pioneering research and various articles by Philippe Grandjean and colleagues at the Harvard School of Public Health Web site at http://www.hsph.harvard.edu/faculty/PhilippeGrandjean.html. Also see the full report on the Inuit in Greenland and the Faroe Islands from the Danish EPA at http://www.mst.dk/homepage/default.asp?Sub=http://www.mst.dk/udgiv/publications/2003/87-7972-477-9/html/kap00_eng.htm.

8. See Theo Colburn, Dianne Dumanoski, and John Peterson Myers, *Our Stolen Future: How We Are Threatening Our Fertility, Intelligence and Survival—A Scientific Detective Story*, foreword by Al Gore (New York: Dutton, 1997) for an account of traveling PCB molecules.

9. The Environmental Health Forum, for example, is a national coalition of some forty health organizations concerned with the environment that is staffed by PSR and shares strategies and a common Web action center, http://www.EnviroHealthAction.org. Much of the E&H movement is found at the grass roots, through coalitions like Lois Gibbs's Center for Health, Environment and Justice at http://www.chej.org; or the Children's Health and Environment Coalition (CHEC) at http://www.checnet.org; and a number of similar groups.

10. See coverage of the EU ban on phthalates at http://www.abc.news.go.com/Business/wireStory?id=909712.

11. Concern about the effects of dioxin go back to the exposure of American GIs (and the Vietnamese who were most heavily exposed) to Agent Orange in Vietnam and to protests over health effects to children in Love Canal lead by Lois Gibbs, then a housewife, now head of CHEJ and author of *Dying for Dioxin: A Citizen's Guide to Reclaiming Our Health and Rebuilding Democracy* (Cambridge, Mass.: South End Press, 1995). Health Care Without Harm (HCWH), a powerful and growing coalition of environmental health groups (of which PSR is a founding member), has built on its initial success in focusing on dioxin produced from medical incinerators. See http://www.noharm.org; and http://www.sfbaypsr.org/work_env_hcwh.htm, for information on PVCs, the phthalate plasticizer, DEHP, and medical waste. This movement is now moving into concern about energy, climate, and green medical buildings.

12. See Robert D. Bullard, *Dumping in Dixie: Race, Class and Environmental Quality* (Boulder, Colo.: Westview Press, 1994); Robert D. Bullard, ed., *Unequal Protection: Environmental Justice and Communities of Color* (New York: Random House, 1994); and Robert D. Bullard et al., eds., *Sprawl City: Race, Politics, and Planning in Atlanta* (Washington, D.C.: Island Press, 2000). These and similar works were inspired by the original work of the United Church of Christ

Commission for Racial Justice, *Toxic Wastes and Race in the United States* (New York: United Church of Christ, 1987).

13. Harris Poll, "Doctors and Teachers Most Trusted among 22 Occupations and Professions: Fewer Adults Trust the President to Tell the Truth," Harris Poll, #61, August 8, 2006. Available at http://www.harrisinteractive.com/harris_poll/index.asp?PID=688.

14. A good brief history of coal restrictions is by David Urbinato, "London's Historic 'Pea-Soupers,' " *EPA Journal* (Summer 1994), available at http://www.epa.gov/history/topics/perspect/london.htm.

15. For a full account of the Donora killer fog and general history of air pollution, see Devra Davis, *When Smoke Ran like Water: Tales of Environmental Deception and the Battle against Pollution* (New York: Basic Books, 2002), chap. 2. For a brief, solid summary of the public health approach to air pollution, see Howard Frumkin, Lawrence Frank, and Richard Jackson, *Urban Sprawl and Public Health: Designing, Planning, and Building for Healthy Communities* (Washington, D.C.: Island Press, 2004), chap. 4.

16. See Clear the Air, *Death, Disease and Dirty Power: Mortality and Health Damage Due to Air Pollution from Power Plants* (September 2000), at http://www.cleartheair.org/fact/mortality/mortalitystudy.vtml. About thirty thousand deaths are from particulates from power plants. Also see Bernie Fischlowitz-Roberts, "Air Pollution Fatalities Now Exceed Traffic Fatalities by 3 to 1," Earth Policy Institute update, September 17, 2002, at http://www.earth-policy.org/Updates/Update17.htm. U.S. figures are estimates, but about half of air pollution deaths (between thirty thousand and thirty-five thousand fatalities) are from auto emissions.

17. See Carol Browner's testimony on the new standards at http://www.resourcescommittee.house.gov/archives/105Cong/fullcomm/Sep30.97/browner.htm. For more information on the NAAQs, see http://www.epa.gov/ttn/oarpg/naaqsfin/naaqs.htm.

18. PSR, "Clinton Strengthens Clean Air Standards," *PSR Reports* 18, no. 2 (Summer 1997), at http://www.psr.org/documents/1997_03.pdf.

19. Richard Wiles and Jacqueline Savitz, "Particle Pollution and Sudden Infant Death Syndrome in the United States," Environmental Working Group/Tides Center and Physicians for Social Responsibility (July 10, 1997), available at http://burningissues.org/pdfs/sids.pdf.

20. PSR Development Department files, reports to foundations.

21. PSR, *A Breath of Fresh Air: How Smarter Energy Choices Can Protect the Health of Pennsylvanians* (September 2004), available from the author.

22. Ibid., 5.

23. Kennedy quoted in Lawrence Wittner, *The Struggle against the Bomb,* vol. 2, *Resisting the Bomb: A History of the World Nuclear Disarmament Movement, 1954–1970* (Stanford, Calif.: Stanford University Press, 1997), 416–423 and ix–x.

24. Lawrence Wittner, *The Struggle against the Bomb,* vol. 3, *Toward Nuclear Abolition: A History of the World Nuclear Disarmament Movement, 1971 to the Present* (Stanford, Calif.: Stanford University Press), 176–177, 195–197, 318–321, 371–373; and David Cortright, *Peace Works: The Citizen's Role in Ending the Cold War* (Boulder, Colo.: Westview Press, 1993).

25. http://www.cleanair-coolplanet.org.

26. For the Harvard Green Campus Initiative supported by Summers, see http://www.greencampus.harvard.edu/about/startup.php. For the original controversy over Summers's remarks on women and science, see Marcella Bombardieri, "Summers' Remarks Draw Fire," *Boston Globe*, June 17, 2005, available at http://www.boston.com/news/local/articles/2005/01/17/summers_remarks_on_women_draw_fire.

27. For background on the sustainability summit, see Peter Bardaglio, "Sustainability Education, the ANAC Mission, and Democratic Citizenship," *ANAC Bulletin* (Fall 2004), at http://www.anac.org/bulletin/archive/Fall04/bul0411–6.html#one. Also see https://www.ithaca.edu/media/release.php?id=1429.

28. PowerPoint presentation by Kent Bransford, MD, PSR Conference on Energy Security. Available from Dr. Bransford through PSR.

29. Carl Pope, remarks at the Symposium on Religion, Science, and the Environment,

Santa Barbara, Calif., November 6–8, 1997, available at http://www.christianecology.org/Carl Pope.html; and a shorter, revised version for the Sierra Club magazine, Carl Pope, "Ways and Means: Reaching beyond Ourselves; It's Time to Recognize Our Allies in the Faith Community," *Sierra* (November/December 1998), at http://www.sierraclub.org/sierra/199811/ways.asp.

30. Author interview with Paul Gorman, NRPE, September 20, 2006.

31. Sir John Houghton, "Climate Change: A Christian Challenge and Opportunity," presentation to the National Association of Evangelicals, Washington, D.C., March 2005, at http://www.creationcare.org/resources/climate/houghton.php.

CHAPTER 5. ASSESSING THE BIG GREENS

1. Jeremy Leggett, *The Carbon War: Global Warming and the End of the Oil Era* (New York: Routledge, 2001), 2–7.

2. Young quoted in Mark Dowie, *Losing Ground: American Environmentalism at the Close of the Twentieth Century* (Cambridge, Mass.: MIT Press, 1996), 25.

3. Dowie, *Losing Ground*, ix. For another influential critique of mainstream environmentalism, see Robert Gottlieb, *Forcing the Spring: The Transformation of the American Environmental Movement* (Washington, D.C.: Island Press, 1993).

4. Dowie, *Losing Ground*, xi–xii.

5. Mary Lou Finley, "Shaping the Movement," in Jonathan Isham and Sissel Waage, eds., *Ignition: What You Can Do to Fight Global Warming and Spark a Movement* (Washington, D.C.: Island Press, 2007), 33–56. Also see Mary Lou Finley et al., *Doing Democracy: The MAP Model for Organizing Social Movements* (Gabriola Island, B.C.: New Society Publishers, 2001).

6. Ross Gelbspan, *The Heat Is On: The Climate Crisis, the Cover-up, the Prescription*, updated ed. (New York: Basic Books, 1998), 197–237.

7. Ibid., 172–173.

8. Ross Gelbspan, *Boiling Point: How Politicians, Big Oil and Coal, Journalists, and Activists Have Fueled the Climate Crisis—and What We Can Do to Avert Disaster* (New York: Basic Books, 2004), ix.

9. Ibid., 130–131.

10. Leggett, *The Carbon War*, 9.

11. Ibid., 23. Leggett's account of the 1990 World Climate Conference in Geneva and the influence of the small island states is at 21–27.

12. Ibid., 42–43. GLOBE was founded in 1990 and now is GLOBE International. See http://www.globeinternational.org.

13. Oppenheimer quoted in Leggett, *The Carbon War*, 74.

14. Lindermann quoted in Leggett, *The Carbon War*, 248.

15. Leggett, *The Carbon War*, 175–176.

16. Clinton quoted in Leggett, *The Carbon War*, 87.

17. Leggett, *The Carbon War*, 98–99.

18. The 1992 presidential election results and analysis can be found at http://en.wikipedia.org/wiki/U.S._presidential_election,_1992. Congressional results for 1992 and 1994 are available from the Office of the Clerk, U.S. House of Representatives, at http://clerk.house.gov/member_info/electionInfo/index.html.

19. Adams quoted in Dowie, *Losing Ground*, 178–180.

20. Dowie, *Losing Ground*, 198–199.

21. The U.S. PIRG employs more than four hundred staff, most active in forty-seven states. See http://www.uspirg.org. The Student PIRGs are active on about one hundred campuses and can be found at http://www.studentpirgs.org.

22. Environmental Defense and Physicians for Social Responsibility, *Putting the Lid on Dioxin*, 1995, is in the PSR records at the Swarthmore Library Peace Collection, Swarthmore College, Swarthmore, Pa.

23. PSR and EWG, *Tap Water Blues*, 1995, available at http://www.psr.org/documents/history.pdf.

24. Quoted in Leggett, *The Carbon War*, 73.

25. Ibid., 174–175.

26. Interview with Paul Gorman, September 20, 2006.

27. Carl Pope, "Ways and Means: Reaching beyond Ourselves; It's Time to Recognize Our Allies in the Faith Community," *Sierra* (November/December 1998), at http://www.sierraclub.org/sierra/199811/ways.asp. Pope's original speech is at http://www.christianecology.org/CarlPope.html. Pope's apology includes the statement that he and others misunderstood a widely read essay from the late 1960s that placed some blame for excessive consumption and other ills on the Judeo-Christian tradition. See Lynn White Jr., "The Historical Roots of Our Ecologic Crisis," *Science* 155 (March 1967): 1,203–1,207, at http://www.sciencemag.org/cgi/reprint/155/3767/1203.pdf.

28. For useful summaries and data on Kyoto, see http://en.wikipedia.org/wiki/Kyoto_Protocol.

29. Philip Shabecoff, *Earth Rising: American Environmentalism in the 21st Century* (Washington, D.C.: Island Press, 2001), 8

30. Ibid., 9.

31. Ibid., 113.

Chapter 6. The New Climate Movement

1. Ross Gelbspan, *Boiling Point: How Politicians, Big Oil and Coal, Journalists, and Activists Have Fueled the Climate Crisis—and What We Can Do to Avert Disaster* (New York: Basic Books, 2004), 40.

2. Jeremy Leggett, *The Carbon War: Global Warming and the End of the Oil Era* (New York: Routledge, 2001), 329.

3. Ibid., 329.

4. http://www.pewclimate.org.

5. See investments from Goldman Sachs and other business indicators from the Pew Center at http://www.pewclimate.org/docUploads/1114_BusinessFinal.pdf.

6. Interlaboratory Working Group, *Scenarios of U.S. Carbon Reduction: Potential Impacts of Energy Technologies by 2010 and Beyond* (Berkeley, Calif.: Lawrence Berkeley National Laboratory; Oak Ridge, Tenn.: Oak Ridge National Laboratory, 1997), LBNL-40533 and ORNL-444, September. Findings are most easily available online in the executive summary at: http://ies.lbl.gov/iespubs/40533%20Chapter%203.pdf; or a project description at http://endues.lbl.gov/Projects/5Lab.html. Known as the Five Labs Study, it was attacked by a GAO study commissioned by opponents of the Kyoto Protocol, Republican Senators Larry Craig, Chuck Hagel, Jesse Helms, and Frank Murkowski. The GAO report was based on the opinions of thirty-one energy industry associations.

7. Joseph Romm, *Cool Companies: How the Best Businesses Boost Profits and Productivity by Cutting Greenhouse Gas Emissions* (Washington, D.C.: Island Press, 1999), ix.

8. Mark Dowie, *Losing Ground: American Environmentalists at the Close of the Twentieth Century* (Cambridge, Mass.: MIT Press, 1996), 180.

9. Lawrence Wittner, *The Struggle against the Bomb,* vol. 3, *Toward Abolition: A History of the World Nuclear Disarmament Movement, 1971 to the Present* (Stanford, Calif.: Stanford University Press, 2003), 461.

10. Wack quoted in Romm, *Cool Companies,* 18.

11. Fay quoted in Romm, *Cool Companies*, 20–25.

12. Romm, *Cool Companies*, 22–23. For an activist account of Shell's corporate contradictions and the impact of protests over Brent Spar, see Leggett, *The Carbon War*, 208–213.

13. Romm, *Cool Companies*, 28–29.

14. Ibid., 131.

15. http://www.exxposeexxon.com.

16. http://www.greenpeace.org/usa/campaigns/global-warming-and-energy/stop-exxonmobil. Also see a consumer campaign from Co-Op America at http://www.coopamerica.org/programs/rs/profile.cfm?id=221.

17. Union of Concerned Scientists, *Smoke, Mirrors and Hot Air: How ExxonMobil Uses Big Tobacco's Tactics to Manufacture Uncertainty on Climate Science* (January 2007), at http://ucsusa.org/assets/documents/global_warming/exxon_report.pdf.

18. http://www.saveourenvironment.org.

19. http://defendersofwildlife.org. Accessed November 1, 2006. Also see "Special Section: Countdown to a Meltdown," *Defenders* 81 (Fall 2006): 8–18.

20. For a brief history of ICCR from the Vietnam War era onward, see http://www.casi.org.nz/ccf/PWolfonethicalinv.htm. Its latest campaign is at http://www.iccr.org/issues/globalwarm/featured.php.

21. http://www.cleanair-coolplanet.org.

22. "Climate: A Crisis Averted" can be viewed at http://www.climatecounts.org/whatis.php.

23. See http://www.nativeenergy.com; and http://www.climatecounts.org.

24. See, for example, the Georgia Air Coalition's Report, "A Call to Action for Clean Air," at http://www.georgiaconservancy.org/News/GAC_Report_2006.pdf.

25. For the Southern Coalition for Clean Energy, see http://www.cleanenergy.org.

26. See the Apollo Alliance at http://www.apolloalliance.org/about_the_alliance.

27. A database of more than 488,000 Christian churches is available at http://list.infousa.com/acl_google.htm?bas_vendor=062250&bas_promotionid=0602RSA. There are more than ninety thousand Baptist churches alone in the United States. Starbucks has 5,668 franchises; see http://www.starbucks.com/aboutus/Company_Factsheet.pdf. McDonald's has 31,800 franchises worldwide; see http://www.answers.com/topic/mcdonald-s.

28. http://en.wikipedia.org/wiki/Demographics_of_the_United_States#Religious_affiliation.

29. See Ralph Weltge, ed., *Ministries to Military Personnel* (New York: United Church of Christ Press, 1973).

30. Carl Pope, "Ways and Means: Reaching beyond Ourselves; It's Time to Recognize Our Allies in the Faith Community," *Sierra* (November/December 1998), at http://:www.sierraclub.org/sierra/199811/ways.asp. Pope's original speech is at http://www.christianecology.org/CarlPope.html.

31. Lynn White Jr., "The Historical Roots of Our Ecologic Crisis," *Science* 155 (March 1967): 1,203–1,207, at http://www.sciencemag.org/cgi/reprint/155/3767/1203.pdf.

32. Pope, "Ways and Means," *Sierra*.

33. The Sierra Club has a formal partnership with faith groups and opportunities for its religious members called "Faith and the Environment," at http://www.sierraclub.org/partnerships/faith.

34. Author's interview with Paul Gorman, September 20, 2006.

35. Ibid.

36. Ibid.

37. Ibid. For a history of the founding of NRPE, see http://www.nrpe.org/whatisthepartnership/history_intro01.htm.

38. Author's interview with Paul Gorman, September 20, 2006.

39. The original appeal from the summit can be found at http://environment.harvard.edu/religion/publications/statements/joint_appeal.html.

40. http://www.ncccusa.org/news/00news48.html.

41. http://www.whatwouldjesusdrive.org. For an example of a local religious protest based on this theme, see http://www.commondreams.org/headlines01/0604–02.htm.

42. http://www.creationcare.org.

43. http://pewforum.org/publications/surveys/postelection.pdf; and http://people-press.org/commentary/pdf/103.pdf, for data and analysis of religion in the 2004 election.

44. http://www.sojo.net.

45. Laurie Goodstein, "Evangelical Leaders Join Global Warming Initiative," *New York Times*, February 8, 2006. The original statement by the Evangelical Climate Initiative, "Climate Change: An Evangelical Call to Action," is at http://www.christiansandclimate.org/statement.

46. Sir John Houghton, "Climate Change: A Christian Challenge and Opportunity," address to the National Association of Evangelicals, March 2005, at http://www.creationcare.org/resources/climate/houghton.php. Also see an interesting longer piece by Houghton at http://www.jri.org.uk/brief/christianchallenge.pdf.

47. Author's interview with Paul Gorman, September 20, 2006.

48. I spoke at the NCC 2007 Advocacy Days at a session devoted to climate change. See the National Council of Churches Eco-Justice Web site for ongoing information and actions, at http://www.nccecojustice.org/index.htm.

49. See religious materials and emphasis in the Earth Day Network Web site at http://www.earthday.net. Accessed May 21, 2007.

50. The Interfaith Climate Walk and a list of combined religious and secular sponsors is at http://www.climatewalk.org.

51. See the Step It Up Web site at http://stepitup2007.org. Also see coverage in the *New York Times*, at http://select.nytimes.com/search/restricted/article?res=F60F16F93B5B0C768DDDAD0894DF404482. For the new "Step It Up2," a year before Election Day 2008, see McKibben's invitation letter at http://stepitup2007.org/#letter. Accessed August 13, 2007.

52. See, for example, McKibben's February 20, 2007, cover story in the *Christian Century*, at http://www.christiancentury.org/article.lasso?id=2978; or one of his sermons, "The Comforting Whirlwind," at http://uuministryforearth.org/sermons/McKibbenSermon.htm.

53. http://www.interfaithworks.org/index.html.

54. http://www.gwipl.org/documents/newsrelease_spotlight.asp.

55. http://www.freetheplanet.org.

56. http://www.freetheplanet.org/ftp.asp?id2=12192.

57. http://www.energyaction.net/main.

58. http://climatechallenge.org.

59. http://www.focusthenation.org.

60. For MASSPIRG, see http://masspirg.org/MA.asp?id2=7335&id3=MA&.

61. The Middlebury conference is at http://www.middlebury.edu/about/newsevents/archive/news_2005/news632423578126773592.htm. It was also covered in the *New York Times* as part of the debate over the future of environmentalism: Felicity Barringer, "Paper Sets Off Debate on Environmentalism's Future," *New York Times*, February 6, 2005. The book growing out of the conference and debate is Jonathan Isham and Sissel Waage, eds., *Ignition: How You Can Fight Global Warming and Spark a Movement* (Washington, D.C.: Island Press, 2007).

62. For campus sustainability initiatives and weekly updates, see http://www.aashe.org/archives/bulletin.php.

63. http://www.nwf.org/campusecology.

64. The Association of American Colleges and Universities is at http://www.aacu.org. See "At Arizona State, Degree Marks New Commitment to Sustainability," *AAC&U News* (October 2006), at http://www.aacu.org/aacu_news/AACUNews06/October06/feature.cfm. The American Association for Sustainability in Higher Education (AASHE) at www.aashe.org also reported on the first degree program at Arizona State. Also see National Council for Science and the Environment (NCSE) at http://ncseonline.org.

65. The Mayor's Climate Protection Agreement is at http://www.seattle.gov/mayor/climate. ICLEI and its Montreal conference are at http://www.iclei.org/index.php?id=1487&tx_ttnews[backPid]=983&tx_ttnews[tt_news]=89&cHash=a5c589f721.

66. Lisa Stiffler, "Mayor Wants to Plant 649,000 Trees," *Seattle Post-Intelligencer*, September 6, 2006.

67. Blaine Harden, "Tree-Planting Drive Seeks to Bring a New Urban Cool: Lower Energy Costs Touted as Benefit," *Washington Post*, September 4, 2006.

68. Ibid.

69. See www.americanforests.org for tree-planting programs. Bush's planting cuts are from Harden, "Tree-Planting Drive."

70. The Sierra Club's Cool Cities program allows citizens to get involved with city climate programs and describes various "Green fleet" initiatives. Houston's is described in Sierra Club, "Cool Cities: Solving Global Warming One City at a Time: Sierra Club's Guide to Local Global Warming Solutions," 6, at http://www.coolcities.us/files/coolcitiesguide.pdf. Mayor White's announcement is at www.houstontx.gov/mayor/press/20050408.html.

71. Anderson quoted in Mark Hertsgaard, "Green Grows Grassroots," *Nation*, July 31, 2006. The Salt Lake City Climate Action Plan is at http://www.slcgreen.com/pages/actionplan.htm.

72. An activist's guide to the Cool Cities Campaign is at http://newjersey.sierraclub.org/ConCom/CoolCities/Cool_Cities_Activist_Toolkit_4-4-0623.PDF.

73. Timothy Gardner, "Northeast States to Act on CO_2 Where Bush Won't," Reuters, August 28, 2006, at http://today/reuters.com/misc/PrinterFriendlyPopup.aspx?type=politicsNews&storyID=20. The RGGI website is: http://rggi.org.

74. Tom Pelton, "Clean Air Law Enacted: Ehrlich Signs Law to Cut Power Plant Emissions," Chesapeake Climate Action Network, April, 7, 2006, at http://www.chesapeakeclimate.org/news/news_detail.cfm?id=81. Maryland joined RGGI when Governor Martin O'Malley took office.

75. Felicity Barringer, "Officials Reach California Deal to Cut Emissions," *New York Times*, August 31, 2006.

76. http://nrdc.org/media/pressreleases/060831c.asp.

77. http://www.allianceforclimateprotection.org.

78. http://www.boxofficemojo.com/movies/?id=inconvenienttruth.htm.

79. http://www.stopglobalwarming.org/default.asp.

80. Alan Riding, "Stars Join Their Voices to Support Live Earth," *New York Times*, July 7, 2007. For the story about Gary Dunham, see http://www.climateprotect.org/aa4. Accessed January 12, 2008.

81. http://1skycampaign.org.

82. http://www.civicyouth.org/PopUps/FactSheets/FS-PresElection04.pdf. Accessed October 30, 2006.

83. For LCV election results and polling data, see http://www.lcv.org.

Chapter 7. Where Do Emissions and Energy Come From?

1. The Energy Information Administration: Official Energy Statistics from the U.S. Government (EIA), provides information on energy for the Department of Energy. See "Energy Basics 101," at http://www.eia.doe.gov/basics/energybasics101.html.

2. Ibid.

3. EIA, "U.S. Carbon Dioxide Emissions from Energy Sources: 2006 Flash Estimates" (May 2007), at http://www.eia.doe.gov/oiaf/1605/flash/pdf/flash.pdf.

4. Jeremy Leggett, *Half Gone: Oil, Gas, Hot Air and the Global Energy Crisis* (London: Portobello, 2006), 21.

5. Cited in ibid., 21.

6. Lovins quoted in Paul Roberts, *The End of Oil: On the Edge of a Perilous New World* (New York: Mariner Books, 2004), 215.

7. Roberts, *The End of Oil*, 1–2.

8. The PSR U.S.-Mexico Border Project studied air quality with an EPA grant described at http://www.epa.gov/ehwg/projects_publications/pdf/retrospective_study_on_pediatric_asthma_and_air_quality.pdf.

9. Julian Darley, *High Noon for Natural Gas: The New Energy Crisis* (White River Junction, Vt.: Chelsea Green, 2004), vii, 10.

10. Peter J. Howe, "Romney Says LNG Security Tightened," *Boston Globe*, January 6, 2005. Also see the Sandia National Laboratories risk analysis at http://www.fossil.energy.gov/programs/oilgas/storage/lng/sandia_lng_1204.pdf; and Darley, *High Noon*, 62–63.

11. Darley, *High Noon*, 2.

12. Lewis Strauss, "Remarks Prepared for Delivery at the Founders' Day Dinner, National Association of Science Writers," September 16, 1954. Quoted in Brice Smith, *Insurmountable Risks: The Dangers of Using Nuclear Power to Combat Global Climate Change* (Takoma Park, Md.: Institute for Energy and Environmental Research Press, 2006), 3.

13. The text of the treaty is at http://www.un.org/events/npt2005/npttreaty.html. For analysis, see the Federation of American Scientists at http://www.fas.org/nuke/control/npt.

14. The MIT report is John Deutch et al., *The Future of Nuclear Power: An Interdisciplinary MIT Study* (Cambridge, Mass.: Massachusetts Institute of Technology, 2003).

15. Richard Garwin and Georges Charpak, *Megawatts and Megatons: A Turning Point in the Nuclear Age?* (New York: Alfred A. Knopf, 2001).

16. See the Institute for Energy and Environmental Research fact sheets, "Uranium: Its Uses and Hazards," at http://www.ieer.org/fctsheet/uranium.html; and "Plutonium: Physical, Nuclear, and Chemical Properties," at http://www.ieer.org/fctsheet/pu-props.html. Accessed July 30, 2007.

17. Helen Caldicott, *Nuclear Power Is Not the Answer* (New York: New Press, 2006), 48–50. Also see Garwin's discussion in *Megawatts and Megatons*, 197–200.

18. Ben Daitz, MD, "A Doctor's Journal: Navajo Miners Battle a Deadly Legacy of Yellow Dust," *New York Times*, May 13, 2003. For a summary and review of the public health literature on uranium mining in New Mexico, see Chris Shuey, "The Navajo Uranium Mining Experience, 2003–1952," Southwest Information Center, Uranium Impact Assessment Program at http://www.sric.org/uranium/navajorirf.html. Dr. John Fogarty's original thesis is available from him through PSR.

19. Smith, *Insurmountable Risks*, 135. There is a vast literature on nuclear nonproliferation. For a good, concise summary of current nonproliferation policies and their pros and cons, see Joseph Cirincione, *Bomb Scare: The History and Future of Nuclear Weapons* (New York: Columbia University Press, 2006), 144.

20. For a brief discussion of the current state of wastes at Hanford, see http://www.hanford.gov/communication/reporter/attachments/WHC/1995/p061495a.pdf. For an overview of military nuclear wastes and radiation in general, see Kenneth Lichtenstein and Ira Helfand, "Radiation and Health: Nuclear Weapons and Nuclear Power," in Eric Chivian et al., *Critical Condition: Human Health and the Environment* (Cambridge, Mass.: MIT Press, 1994).

21. Smith, *Insurmountable Risks*, 141.

22. Garwin and Charpak, *Megawatts and Megatons*, 200.

23. See David Robie, *Eyes of Fire: The Last Voyage of the Rainbow Warrior, Memorial Edition* (Auckland: Asia Pacific Network, 2005). Available from South Pacific Books at http://www.southpacificbooks.co.nz. Also see the Rainbow Warrior Web site from Greenpeace New Zealand at http://www.rainbow-warrior.org.nz/home.asp.

24. For an excellent overview of the public health approach to the environment, see Michael McCally, "Environment, Health, and Risk," in Michael McCally, ed., *Life Support: The Environment and Human Health* (Cambridge, Mass.: MIT Press, 2002), 1–14.

25. Smith, *Insurmountable Risks*, 32–33.

26. Garwin and Charpak, *Megawatts and Megatons*, 172–174.

27. Ibid., 175–176.

28. Ibid., 201

29. Smith, *Insurmountable Risks,* 220–222.

30. Ibid., 222–223.

31. Ibid., 222.

32. After the study was revealed by UCS, it was included in House hearings by Rep. Ed Markey (D-MA). U.S. House of Representatives, Committee on Interior and Insular Affairs, Subcommittee on Oversight and Investigation, "Calculation of Reactor Accident Consequences (CRAC-2) for U.S. Nuclear Power Plants," November 1, 1982. The *Washington Post* also reported on the UCS discovery that day. Milton R. Benjamin, "Nuclear Study Raises Estimates of Accident Tolls," *Wash-*

ington Post, November 1, 1982. For warnings of nuclear dangers based on CRAC-2 after 9/11, see, for example, Jim Riccio, *Risky Business: The Probability and Consequences of a Nuclear Accident* (Washington D.C.: Greenpeace, n.d.) at http://www.greenpeace.org/raw/content/usa/press/reports/risky-business-the-probabilit.pdf.

33. National Research Council, Committee to Assess Health Risks from Exposure to Low Levels of Ionizing Radiation, *Health Risks of Exposure to Low Levels of Ionizing Radiation: BEIR VII, Phase 2* (Washington, D.C.: National Academies Press, 2006), at http://newton.nap.edu/books/030909156X/html/R1.html. Quoted in Smith, *Insurmountable Risks*, 227.

34. For a comprehensive look at concerns from critics of the Bush administration that not enough was being done to protect nuclear power plants from attack and legislation attempting to remedy the situation, see the 2004 report by the Congressional Research Service (CRS), "Nuclear Power Plants: Vulnerability to Terrorist Attack," on the Federation of American Scientists Web site at http://www.fas.org/irp/crs/RS21131.pdf. Also see the 2002 report, "Nuclear Power Plant Security: The View from inside the Fences," which used extensive interviews with plant guards and personnel by the Project on Government Oversight (POGO), at http://www.pogo.org/p/environment/eo-020901-nukepower.html.

35. See Lisbeth Grunland, David Lochbaum, and Edwin Lyman, *Nuclear Power in a Warming World: Assessing the Risks, Addressing the Challenges* (Cambridge, Mass.: Union of Concerned Scientists, 2007), for a critique of nuclear power policy and planning as of 2008.

36. The PSR report is at http://www.psr.org/site/DocServer/PSR_NuclearTerr_rpt_full.pdf?docID=781.

37. Figures are from a strong case by two leading realists about coal. See Ken Berlin and Robert M. Sussman, *Global Warming and the Future of Coal: The Path to Carbon Capture and Storage* (Washington, D.C.: Center for American Progress, 2007), 1.

38. Ibid., 6.

39. The EFC report and various participants and funders can be found at http://www.energyfuturecoalition.org/pubs/EFC%20Report.pdf. Accessed July 30, 2007.

40. National Commission on Energy Policy, *Ending the Energy Stalemate: A Bipartisan Strategy to Meet America's Energy Challenges* (Washington, D.C., 2004), at http://www.energycommission.org/files/contentFiles/report_noninteractive_44566feaabc5d.pdf. Discussion of coal is at 51–56.

41. See Gore's energy proposals in his detailed speech at NYU Law School at http://www.nyu.edu/community/gore.html. He only endorses new coal technologies linked to a carbon tax and without CO_2 emissions.

42. See the late 2005 announcement by DOE Secretary Samuel Bodman at http://fossil.energy.gov/news/techlines/2005/tl_futuregen_signing.html.

43. Hammerschlag quoted in Craig Canine, "How to Clean Coal," *OnEarth* (Fall 2005), at http://www.nrdc.org/onearth/05fal/coal1.asp.

44. Robert F. Service, "The Carbon Conundrum," *Science* 305 (2004): 962–963.

45. See the Department of Energy's 2005 Carbon Sequestration Programmatic Environmental Impact Statement: alternatives including proposed action. DOE/EIS-0366. Also see information at http://www.netl.doe.gov/publications/carbon_seq/Program064_4P.pdf.

46. DOE and an overview by Victor E. Camp, Geological Sciences Department, San Diego State University, at http://www.geology.sdsu.edu/how_volcanoes_work/Nyos.html. Also see a Reuters story on MSNBC.com, "Bury Gas Tied to Warming? Idea Has Its Risks," August 3, 2006, at http://www.msnbc.com/id/13803974/print/1/displaymode/1098. Accessed August 4, 2006.

47. See a review of the study in *Environmental News and Technology* at http://pubs.acs.org/subscribe/journals/esthag-w/2006/sep/tech/rr_co2.html. The full scientific article, Y. K. Kharaka et al., "Gas-Water-Rock Interactions in Frio Formation Following CO_2 Injection: Implications for the Storage of Greenhouse Gases in Sedimentary Basins," *Geology* 34, no. 7 (July 26, 2006): 577–580, is at http://www.gsajournals.org/archive/0091–7613/34/7/pdf/i0091–7613–34–7–577.pdf. Accessed July 31, 2007. For a balanced overview, see Carola Hanisch, "The Pros and Cons of

Carbon Dumping," *Environmental Science and Technology* (January 1998), at http://pubs.acs.org/hotartcl/est/98/jan/carbon.html.

48. Carrie LaSeur, "The Merchant of Menace: How 'Merchant Coal' Is Changing the Face of America," *Grist*, August 24, 2006, at http://www.grist.org/cgi-bin/printthis.pl. Accessed September 20, 2006.

49. Ibid.

50. Canine, "How to Clean Coal."

51. Ibid.

52. Jeff Goodell, *Big Coal: The Dirty Secret Behind America's Energy Future* (New York: Houghton Mifflin, 2006), 223.

CHAPTER 8. ENERGY FUTURES

1. Paul Roberts, *The End of Oil: On the Edge of a Perilous New World* (New York: Mariner Books, 2004), 326.

2. Ibid., 31–35. Also see http://www.aaca.org/history/chronological/cars_30.htm. Americans drove 198 billion miles as early as 1929. Also Chris Flavin, lecture and Worldwatch Institute PowerPoint presentation, "Renewable Energy at the Tipping Point," American University, March 2006.

3. John Vidal, "Sweden Plans to Be World's First Oil-Free Economy," *Guardian*, February 8, 2006. Accessed at http://www.guardian.co.uk/oil/story/0,,1704954,00.html#article_continue.

4. Flavin, "Renewable Energy at the Tipping Point."

5. For statistics on wind power, see the American Wind Energy Association at http://www.awea.org. GE investments are in Greenbiz.com, "General Electric Announces Major Wind Power Investments," February 16, 2007, at http://www.greenbiz.com/news/news_third.cfm?NewsID=34604.

6. Flavin, "Renewable Energy at the Tipping Point."

7. Ibid.

8. Rob Taylor, "Australian First Wave Power Plant Ready to Roll," Reuters, March 1, 2007. Available at the Environmental News Network, at http://enn.com/today.html?id=12314. For geothermal energy, see the Union of Concerned Scientists at http://www.ucsusa.org/clean_energy/renewable_energy_basics/offmen-how-geothermal-energy-works.html. Also, "The Future of Geothermal Energy: Impacts of Enhanced Geothermal Systems [EGS] on the United States in the 21st Century" (MIT, 2006), at htpp://www.geothermal.inel.gov.

9. John Ritter, "Wind Turbines Taking Toll on Birds of Prey," *USA Today*, January 1, 2004.

10. Mick Sagrillo, "Advice from an Expert," American Wind Energy Association, at http://www.awea.org/faq/sagrillo/swbirds.html#17.

11. John Flicker, "Wind Power," *Audubon* (November-December 2006), at http://www.audubon.org/campaign/windPowerQA.html.

12. See "Mass Audubon's Position on the Cape Wind Energy Project," at http://www.massaudubon.org/printnews.php?id=317&type=wind. Accessed August 7, 2007. For the full fascinating conflict, see Wendy Williams and Robert Whitcomb, *Cape Wind: Money, Celebrity, Class, Politics, and the Battle for Our Energy Future* (New York: Public Affairs, 2007). See the attacks on Williams for supporting Cape Wind in Wendy Williams, "A Mighty Wind: The News They Don't Want to Hear," *E: The Environmental Magazine* 17, no. 4 (July/August 2007) at http://www.emagazine.com/view/?3824.

13. Amandeep Dhanju, Phillip Whitaker, and Sandra Burton, "Assessment of Delaware Offshore Wind Power," College of Marine Studies, University of Delaware (September 2005), at http://www.ocean.udel/windpower/docs/BurDanWhit05-MAST-667-FINAL.pdf. Also see "Survey Shows Strong Support for Offshore Wind Power," *UD Daily*, University of Delaware, January 16, 2007, at http://www.udel.edu/PR/UDaily/2007/jan/wind011607.html.

14. Worldwatch Institute and the Center for American Progress, *American Energy: The Renewable Path to Energy Security* (Washington, D.C., 2007), 26–27, at http://www.worldwatch.org/node/4405.

15. Fuel economy statistics are available from the U.S. Department of Transportation at

http://www.nhtsa.dot.gov/staticfiles/DOT/NHTSA/Vehicle%20Safety/Articles/Associated%20Files/SummaryFuelEconomyPerformance-2005.pdf.

16. Tom Krisher, "Toyota and DaimlerChrysler Pass Ford," AP, Washington Post.com, February 1, 2007, at http://www.washingtonpost.com/wp-dyn/content/article/2007/02/01/AR2007020101534_pf.html. For information about tax credits for Prius buyers and the sixty thousand–vehicle eligibility limit per auto manufacturer, see http://www.toyota.com/prius/tax.html.

17. Sherry Boschert, *Plug-in Hybrids: The Cars That Will Recharge America* (Gabriola Island, B.C.: New Society Publishers, 2007), 30–31.

18. California Air Resources Board, *The California Almanac of Emissions and Air Quality, 2005 Edition*, 90–91, at http://www.arb.ca.gov/aqd/almanac/almanac05/almanac2005all.pdf.

19. "GM: Hydrogen Cars Will Re-establish Company," MSNBC, July 15, 2006.

20. Worldwatch Institute and the German Agency for Technical Cooperation, "Biofuels for Transportation: Global Potential and Implications for Sustainable Agriculture and Energy in the 21st Century" (June 2006), summary, 1–2. The full version of the report can be downloaded at: http://www.worldwatch.org/pubs/biofuels.

21. Ibid., 3.

22. Ibid., 1. Also see Matthew L. Wald, "G.M. Buys Stake in Ethanol Made from Waste," *New York Times*, January 14, 2008.

23. From a PowerPoint presentation by Vinod Khosla, "Biofuels: Think outside the Barrel," at the PowerShift annual conference, July 2006, hosted by the energy security group 20/20 Vision. Its Web site is at http://www.2020Vision.org. The presentation is available from Khosla at vk@khoslaventures.com.

24. Ibid.

25. Ibid.

26. Feldman quoted in Daniel Howden, "The Big Green Fuel Lie," *Independent*, March 5, 2007, at http://news.independent.co.uk/environment/climate_change/article2328821.ece.

27. Sabrina Valle, "Losing Forests to Fuel Cars," *Washington Post*, July 31, 2007.

28. *Who Killed the Electric Car?*, directed by Chris Paine (Sony Pictures, 2006).

29. Boschert, *Plug-in Hybrids*, 32. Also see Michiline Maynard, "Toyota Will Offer a Plug-in Hybrid by 2010," *New York Times*, January 14, 2008.

30. Stephen Foley, "U.S. Car Giants Launch Green Drive," *Independent*, January 8, 2007, at http://news.independent.co.uk/business/news/article2134926.ece.

31. Amory B. Lovins et al., *Winning the Oil Endgame: Innovation for Profit, Jobs, and Security* (Snowmass, Colo.: Rocky Mountain Institute, 2004).

32. Ron Gluckman, "Is It a Bird, Is It a Plane?" *Asian Wall Street Journal*, April 21–23, 2003, at http://www.ucsusa.org/clean_energy/renewable_energy_basics/offmen-how-geothermal-energy-works.html.

33. Jack Doyle, *Taken for a Ride: Detroit's Big Three and the Politics of Air Pollution* (New York: Four Walls, Eight Windows, 2000).

34. See the Web site of We Act, a Harlem environmental justice group, at http://www.weact.org/index.html. Also see Frederica P. Perera et al., "The Challenge of Preventing Environmentally Related Diseases in Young Children: Community-Based Research in New York City," *Environmental Health Perspectives* 110 (2002): 197–204.

35. Howard Frumkin, Lawrence Frank, and Richard Jackson, *Urban Sprawl and Public Health: Designing, Planning, and Building for Healthy Communities* (Washington, D.C.: Island Press, 2004), 39–41.

36. Frumkin, Frank, and Jackson, *Urban Sprawl*, box 4–2.

37. "Slow Growth Wins," *Washington Post*, November 12, 2006.

38. Frumkin, Frank, and Jackson, *Urban Sprawl*, 70, box 4–1.

39. Robert D. Bullard, Glenn S. Johnson, and Angel O. Torres, *Sprawl City: Race, Politics, and Planning in Atlanta* (Washington, D.C.: Island Press, 2000), x.

40. The home page for William McDonough and Partners is at http://www.mcdonoughpartners.com.

41. *The Next Industrial Revolution: William McDonough, Michael Braungart, and the Birth of the Sustainable Economy*, narrated by Susan Sarandon (Earthome Productions, 2001), is available at http://thenextindustrialrevolution.org. Also see McDonough's pathbreaking book, made from entirely recycled, nonpaper, waterproof materials, William McDonough and Michael Braungart, *Cradle to Cradle: Remaking the Way We Make Things* (New York: North Point Press, 2002).

42. The home page for the Adam Joseph Lewis Center for Environmental Studies is at http://www.oberlin.edu/ajlc/ajlcHome.html. For background on David Orr and his role in promoting sustainability, see http://www.oberlin.edu/news-info/98sep/orr_profile.html.

43. See the Condé Nast building, completed in 1999, at http://www.nyc-architecture.com/MID/MID109.htm. The Bank of America Tower, begun in 2004 and scheduled to open in 2008, can be seen at http://www.nyc-architecture.com/MID/MID157.htm. Also see Steven Mufson, "As Power Bills Soar, Companies Embrace 'Green' Buildings," *Washington Post*, August 5, 2006.

44. John Ritter, "Building 'Green' Reaches a New Level," *USA Today*, July 26, 2006.

45. Ibid.

46. NRDC, "Affordable Green Housing," at http://nrdc.org/cities/building/fhousing.asp. Accessed September 15, 2006.

47. Ritter, *USA Today*, July 26, 2006.

48. Health Care Without Harm's web site is: http://www.noharm.org/

49. Ritter, "Building 'Green' Reaches a New Level."

50. See the Emory facility at http://www.whsc.emory.edu/_pubs/em/2005spring/gifts_support.html.

51. For a variety of views on President Bush's first energy proposal, see the PBS Newshour, May 23, 2001, at http://www.pbs.org/newshour/bb/environment/jan-june01/fossil_5–23.html. For an overall environmentalist assessment of President Bush's first year, including energy, see Matthew Gross, "Beating around the Bush," *Grist*, February 12, 2002, at http://www.grist.org/comments/soapbox/2002/02/12/gross-bush.

52. John M. Broder, "Bush Signs Broad Energy Bill," *New York Times*, December 19, 2007. For the reaction of the environmental community and the need to improve Congress and the White House, see the League of Conservation Voters at http://www.lcv.org/newsroom/press-releases/lcv-praises-house-passage-of-energy-bill-as-good-first-step.html.

53. National Commission on Energy Policy, *Ending the Energy Stalemate: A Bipartisan Strategy to Meet America's Energy Challenges* (Washington, D.C., 2004), at http://www.energycommission.org/files/contentFiles/report_noninteractive_44566feaabc5d.pdf.

54. Ibid., xiv.

55. David J. Lynch, "ExxonMobil Amasses Record $36B 2005 Profit," *USA Today*, January 30, 2006. Also, John Porretto, "ExxonMobil Posts Record Annual Profit," AP, February 1, 2007, at http://www.breitbart.com/article.php?id=D8N0VRD80&show_article=1. ExxonMobil made $36.13 billion in 2005 and broke its own record with $39.5 billion in 2006.

56. National Commission on Energy Policy, *Ending the Energy Stalemate,* v. This is recommendation no. 4 of the commission. Nuclear power, for example, gets $2 billion over ten years.

57. Charles F. Kutscher, ed., *Tackling Climate Change in the U.S.: Potential Carbon Emissions Reductions from Energy Efficiency and Renewable Energy by 2030* (American Solar Energy Society, 2007). The report can be downloaded at http://www.ases.org/climatechange.

58. Andrew C. Revkin, "Budgets Falling in Race to Fight Global Warming," *New York Times,* October 30, 2006.

59. Kutscher, "Executive Summary," in *Tackling Climate Change in the U.S.*, 5.

60. Ibid., 4.

61. See the Voice of America report on the release of the ASES Report to Congress at http://www.voanews.com/english/archive/2007–02/2007–02–02-voa57.cfm?CFID=61175188&CFTOKEN=18562975.

62. The action link is available at http://www.americanenergynow.org.

63. Information on *The Energy (R) Evolution* is available at the EREC Web site at http://www.erec-renewables.org/publications/EREC_publications.htm#revolution.

64. See Arjun Makhijani, *Carbon-Free and Nuclear-Free: A Roadmap for U.S. Energy Policy* (Takoma Park, MD: Institute for Energy and Environmental Research Press and RDR Books, 2007). This authoritative and useful plan was released at the very end of 2007, too late to be discussed in this book. An executive summary and the free newsletter *Science for Democratic Action* can be had at http://www.ieer.org.

CHAPTER 9. CREATING HOPE

1. Laurie David, *The Solution Is You! An Activist's Guide* (Golden, Colo.: Speaker's Corner Books, 2006), 18.

2. Ibid., 20.

3. Keith Bradsher, *High and Mighty: The Dangerous Rise of the SUV*, reprint ed. (New York: Public Affairs, 2004).

4. David, *The Solution Is You!*, 34.

5. Jeffrey Langholz and Kelly Turner, *You Can Prevent Global Warming (and Save Money!): 51 Easy Ways* (Kansas City, Mo.: Andrews McMeel, 2003), 3–7.

6. Ibid., 9.

7. Ibid., 10–11.

8. Ibid., 24–25.

9. David Steinman, *Safe Trip to Eden: Ten Steps to Save Planet Earth from the Global Warming Meltdown*, foreword by Wendy Gordon Rockefeller (New York: Thunder's Mouth Press, 2007), 175–176.

10. Langholz and Turner, *You Can Prevent*, 185–186.

11. For tree-planting programs, see the National Arbor Day Foundation at http://arborday.org; and American Forests at http://www.americanforests.org.

12. Information on the Enterprise partnership is at http://arborday.org/eneterprise. Accessed October 24, 2006.

13. Information on Flexcar is at http://www.flexcar.org. Zipcar is at http://www.zipcar.com. Also see Sara Kehaulani Goo, "Car-Sharing Merges into the Mainstream," *Washington Post*, September 5, 2006.

14. Dan Chiras, *The Homeowner's Guide to Renewable Energy: Achieving Energy Independence through Solar, Wind, Biomass and Hydropower* (Gabriola Island, B.C.: New Society Publishers, 2006). Also see Scott Sklar, *Consumer Guide to Solar Energy*, 3rd ed. (Chicago: Bonus Books, 2002).

15. Chiras, *The Homeowner's Guide*, 3.

16. Ibid., 4.

17. Ibid.

18. The story on Enertia homes is at http://www.greenbiz.com/news/printer.cfm?NewsID=33480. Accessed August 4, 2006. Information on green buildings and the growing industry is at http://www.GreenerBuildings.com.

19. See Barbara Kingsolver, with Steven L. Hopp and Camille Kingsolver, *Animal, Vegetable, Miracle: A Year of Food Life* (New York: HarperCollins, 2007); and Michael Pollan, *The Omnivore's Dilemma: A Natural History of Four Meals* (New York: Penguin Books, 2007); as well as Langholz and Turner, *You Can Prevent*, 252.

20. Langholz and Turner, *You Can Prevent*, 251.

21. Ibid., 252.

22. Ibid., 257.

23. Ibid., 260.

24. http://www.american.edu/academic.depts/sis/campaignforsis/designplans.html.

25. See details of the Harvard Green Campus Initiative at http://www.greencampus.harvard.edu.

26. Langholz and Turner, *You Can Prevent*, 230–231.

27. Ibid.

28. Ibid., 234–235. Carbon offset companies are at http://www.climatepartners.com; and http://www.tripleE.com. The nonprofit CarbonFund is at http://www.carbonfund.org.

29. Elysa Hammond's presentation was at Middlebury College, September 30, 2006.

30. Cool Tags are at http://www.clifbar.com. NativeEnergy is at http://www.nativeenergy.com.

31. For the Expedia initiative with TerraPass, see http://www.greenbiz.com/news/printer.cfmNewsID=33852. Accessed on October 21, 2006. TerraPass is at http://www.terrapass.com.

32. Robert Putnam, *Bowling Alone: The Collapse and Revival of American Community*, new ed. (New York: Simon and Schuster, 2001).

33. Energy Information Administration, "Emissions of Greenhouse Gases in the United States 2005," at http://www.eia.doe.gov/oiaf/1605/ggrpt/carbon.html. Accessed May 28, 2007.

34. Saul Bellow, *Herzog* (New York: Viking Adult, 1964).

Chapter 10. Hope for a Heated Planet

1. Interview with Paul Gorman, National Religious Partnership for the Environment, September 20, 2006.

2. For Harvard's Green Campus Initiative, see http://www.greencampus.harvard.edu.

3. For ExxonMobil's 2005 profits, see http://www.ExxonMobil.com/Corporate/Files/Corporate/fo_2005.pdf. and www.msnbc.msn.com/id/11098458. The company's third-quarter 2006 earnings are at http://money.cnn.com/2006/07/27/news/companies/exxon. For critical views of ExxonMobil and shareholder challenges, see http://www.campaignexxonmobil.org; and http://www.stopexxonmobil.org.

4. Roy Morris Jr., *Fraud of the Century: Rutherford B. Hayes, Samuel Tilden, and the Stolen Election of 1876* (New York: Simon and Schuster, 2003).

5. "Redefining Political Attitudes and Activism," Harvard Institute of Politics, November, 16, 2005, at http://www.iop.harvard.edu/pdfs/survey/fall_2005_execsumm.pdf; and http://www.lcv.org/campaigns/2004-presidential. With 59,028,109, Kerry received over 8 million more popular votes than Gore did. See http://www.cnn.com/Election/2000 and http://www.infoplease.com/ipa/A0922901.html.

6. Joe Klein mentioned Nixon's and Colson's attacks in "The Long War of John Kerry: Can a Massachusetts Brahmin Become President?" *New Yorker*, December 12, 2002. For review of the media handling of this issue, also see http://mediamatters.org/items/200405040004; and http://dir.salon.com/story/opinions/conason/2004/04/23/o_neill/index.html.

7. http://www.lcv.org.

8. http://www.sierraclub.org; and undated mailing, Maryland Sierra Club, October 2006.

9. *Washington Post*, editorial, October 25, 2006, and candidate endorsements, November 7, 2006.

10. "Campaign 2006," *Greenwire*, November 1, 2006. These are estimates and are not a full accounting of what environmental groups will have spent in 2006.

11. http://sequestration.mit.edu/research/survey2006.html.

12. http://www.trumanlibrary.org/9981.htm.

13. http://www.army.mil/cmh-pg/books/integration/IAF-22.htm.

14. World War II production figures are from the National Defense University's Institute for National Strategic Studies at http://www.ndu.edu/inss/McNair/mcnair50/m50c9.html.

15. For a white paper and facts about limited U.S. efforts in diplomacy and aid, see Coalition for U.S. Leadership Abroad, at http://www.colead.org; and a bipartisan campaign to improve U.S. engagement and spending on diplomacy at the U.S. Global Leadership Campaign, at http://www.usgloballeadership.org.

16. Andrew C. Revkin, "Budgets Falling in Race to Fight Global Warming," *New York Times*, October 30, 2006.

17. See analysis by the Center for American Progress, "Bush's Energy Budget: Proposals Not Consistent with Claims," February 8, 2008, at http://www.americanprogress.org/issues/2008/02/energy_budget.html.

18. Daniel M. Kammen and Gregory F. Nemet, "Real Numbers: Reversing the Incredible Shrinking Energy R&D Budget," *Issues in Science and Technology* (Fall 2005): 84, at http://www.issues.org/22.1/realnumbers.html.

19. Daniel M. Kammen, "Climate Change Technology Research: Do We Need a 'Manhattan Project' for the Environment?," testimony at hearings, U.S. House of Representatives, Committee on Government Reform, September 21, 2006, at http://reform.house.gov/GovReform/Hearings/EventSingle.aspx?EventID=50466.

20. *Washington Post*, October 30, 2006, carried a Reuters story by Adrian Croft and Gerard Wynn that focused on the "apocalyptic" results of climate inaction; see http://www.washingtonpost.com/wp-dyn/content/article/2006/10/30/AR2006103000258_pf.html. Other stories, and the reaction of the liberal Center for American Progress, which called the report "a bleak picture," were similar. See http://www.americanprogressaction.org/atf/cf/%7B65464111-BB20–4C7D-B1C9–0B033.

21. Stern's comments are from the HM Treasury press release, "Publication of the Stern Review on the Economics of Climate Change," October 30, 2006, at http://www.hm-treasury.gov.uk/newsroom_and_speeches/press/2006/press_stern06.cfm. Accessed October 31, 2006. The full Stern Commission Report can be downloaded at http://www.hm-treasury.gov.uk/Independent_Reviews/stern_review_economics_climate_change/sternreview_index.cfm.

22. http://www.solar2006.org/dispatches/solution.html.

23. Greenpeace and the European Photovoltaics Industry Association (EPIA), *Solar Generation: Solar Electricity for Over 1 Billion People and 2 Million Jobs by 2020* (September 2006), at http://www.greenpeace.org/raw/content/international/press/reports/solar-generation-ii.pdf.

24. Apollo Strategy Center, *New Energy for Cities: Energy-Saving and Job Creation Policies for Local Governments*, (Washington, D.C.: Apollo Alliance, 2006), at http://www.apolloalliance.org/downloads/resources_new_energy_cities.pdf.

25. Robert Socolow et al., "Solving the Climate Problem: Technologies Available to Curb CO_2 Emissions," *Environment* 46, no. 10 (December, 2004): 8–19. Originally in *Science* 305, no. 5686 (August 13, 2004): 968–972.

26. Ibid., 10.

27. Ibid., 11.

28. Ibid., 12.

29. Oliver Bullough, "Home Wind Turbines Turn Fashionable in Britain," Reuters, October 12, 2006, at http://www.enn.com/today_PF.html?id=11435. Accessed October 12, 2006.

30. Ibid. (Figures are given in both U.S. dollars and pounds sterling.) Also see Robert Booth, "Cameron Plans Wind Turbine for His Roof," *Guardian*, March 4, 2006.

31. Socolow et al., "Solving the Climate Problem," 15–16.

32. The U.S. PIRG report, "Rising to the Challenge: Six Steps to Cut Global Warming Pollution in the United States" (August 2006), is at http://uspirg.org/uspirg.asp?id2=26147. Accessed September 20, 2006. The 2006 NRDC report, "A Responsible Energy Plan for the United States," is at http://nrdc.org/air/energy/rep/execsum.asp. Accessed October 10, 2006.

33. For an analysis and a sharp critique, see materials from Citizens against Government Waste, at http://www.cagw.org/site/News2?page=NewsArticle&id=8684.

Index

About the Author

Robert K. Musil, Ph.D., M.P.H., is a scholar-in-residence and adjunct professor in the School of International Studies at American University where he teaches in the Program on Global Environmental Politics and the Nuclear Studies Institute. He is the past executive director and CEO of the Nobel Peace Prize–winning Physicians for Social Responsibility (PSR) and served from 1992 to 2006. He was the executive producer and host of "Consider the Alternatives," a weekly radio program syndicated to over 150 stations with 2 million listeners and a two-time winner of the Armstrong Award for Excellence in Radio Broadcasting. Dr. Musil is also a popular campus lecturer and Woodrow Wilson Foundation Visiting Fellow, and a visiting scholar at the Churches' Center for Theology and Public Policy at Wesley Seminary, where he teaches religious perspectives on climate change and security. He is chairman of the board of 2020 Vision, Environment and Energy Solutions.